# Climate Health Risks in Megacities

## Sustainable Management and Strategic Planning

# Climate Health Risks in Megacities

## Sustainable Management and Strategic Planning

## Cesar Marolla

Foreword by Dr. Alfred Sommer

CRC Press
Taylor & Francis Group
Boca Raton London New York

CRC Press is an imprint of the
Taylor & Francis Group, an **informa** business

CRC Press
Taylor & Francis Group
6000 Broken Sound Parkway NW, Suite 300
Boca Raton, FL 33487-2742

First issued in paperback 2019

© 2017 by Taylor & Francis Group, LLC
CRC Press is an imprint of Taylor & Francis Group, an Informa business

No claim to original U.S. Government works

ISBN-13: 978-1-4987-6774-3 (hbk)
ISBN-13: 978-0-367-87815-3 (pbk)

---

**Library of Congress Cataloging-in-Publication Data**

---

Names: Marolla, Cesar, author.
Title: Climate health risks in megacities : sustainable management and strategic planning / Cesar Marolla.
Description: Boca Raton, FL : CRC Press/Taylor & Francis Group, [2016] | Includes bibliographical references and index.
Identifiers: LCCN 2016027864| ISBN 9781498767743 (hardback : acid-free paper) | ISBN 9781315367323 (ebook)
Subjects: LCSH: Urban health. | Climatic changes--Health aspects. | Sustainable urban development. | City planning--Health aspects. | Emergency management--Planning. | Urban ecology (Sociology)
Classification: LCC RA566.7 .M368 2016 | DDC 362.1/042--dc23
LC record available at https://lccn.loc.gov/2016027864

---

Visit the Taylor & Francis Web site at
http://www.taylorandfrancis.com

and the CRC Press Web site at
http://www.crcpress.com

*First, my deepest appreciation and love to my parents and my wife, Lisa, for*
*their love and invaluable support. Also to my three boys—Nicholas, Alexander,*
*and Michael—for the early morning fights before school and the evenings full of*
*arguments, noise, laughter, and video games that kept me from finishing this book*
*in a timely manner. I will isolate myself from the rest of the world for my next book!*

*Furthermore, I would like to dedicate this book to my former Harvard professor,*
*Dr. Robert Pojasek, whose leadership, support, attention to detail, hard work,*
*and scholarship have set an example I hope to match someday. His mentorship*
*was vital to my personal and professional growth, and I am eternally indebted.*

*Dr. Alfred Sommer and Dr. Yuval Neria taught me through their determination*
*and passion to follow my heart and the value of pushing myself beyond*
*what I thought were my limits. Their inspiring and important work, which*
*has had a major impact on people's lives, have shown me that commitment*
*and courage can conquer the most difficult challenges. Thank you!*

*A special dedication is due to the late physician and associate director of the Center*
*for Health and the Global Environment at Harvard Medical School, Dr. Paul*
*Epstein. His expertise, commitment, and dedication to global health and the link*
*between the spread of infectious diseases and extreme weather events have opened*
*new doors leading to new questions and gaining new insights. His pioneer work*
*added a new dimension to research into the potential impacts of climate change.*

*Finally, thank you to all the men and women in uniform*
*who serve and protect our rights and freedoms.*

# Contents

# Section II   Risk Management and Its Relation to Climate Change

**Section III   Global Health Risks, Facts, and
Challenges of Climate Change Implication**

## Section IV  Case Study: Researched Megacities

## Section V   Prognosis for Change

# Foreword

The magnitude and consequences of climate change have largely been buried in global estimates, where the health impact in some areas may cancel that in others. A recent, carefully constructed exercise developed a statistically likely range of estimates of the likely impact climate change would have, over varying points in time, in discrete regions of the United States. It therefore provides a level of granularity that few previous studies could, while at the same time, providing statistically linked likelihoods of the probability of the degree of change(s) that will be encountered (*Risky Business: The Economic Risks of Climate Change in the United States; A Climate Risk Assessment for the United States*, June 2014). Although this exercise was primarily focused on the economic consequences of climate change over the mid to longer terms, it portends a dramatic impact on the health of Americans (and by extension, people around the world).

Whereas some consequences of climate change, like greater numbers of storms and other extreme weather events (e.g., Hurricane Sandy) can be anticipated, their severity and timing cannot. But recent experience shows the dramatic impact they can have, with flooding resulting in large numbers of homeless (in some areas of New York displaced families are yet to be permanently resettled), closure of hospitals (NYU Langone Medical Center lost all power, forcing it to evacuate, and find appropriately equipped and staffed sophisticated professional facilities—neonatal intensive care units—for nearly 100 vulnerable, premature infants), and the rescue and treatment of many displaced people. But one, clearly predictable, persistent, and growing consequence will be a dramatic rise in temperature in many heavily populated areas, some of which constitute much of the countries present "bread basket."

The Risky Business climate models predict, with reasonable certainty, a dramatic rise in extreme temperatures affecting much of the Southwest, Southeast, and lower Midwest. We know what that single, predictable change will mean for the public's health. Within 35 years the average summer in Montana will be hotter than it presently is in New Mexico. By the end of the century, summers in Oregon, Washington, and Idaho will be hotter than they are today in Texas! Of course wealthy Americans can ignore heat waves by staying inside air-conditioned facilities (assuming that power grids can handle the electrical demands). But many people will not be able to afford air-conditioning, or must work out of doors in construction, agriculture, and similar industries. It is not just that they will be uncomfortable; they may well die. A "wet-bulb" temperature (a measurement that employs the impact of both heat and humidity) above 95 degrees Fahrenheit is incompatible with long-term survival.

As temperatures increase we sweat; evaporation of that sweat lowers our surface temperature, helping to keep our core body temperature at a safe level. But when humidity is high, sweat cannot evaporate, and we become incapable of maintaining our core temperature within a range compatible with survival. This is true for healthy young individuals but even more so for the elderly and frail. All those who can't afford to remain in air-conditioned facilities, or don't have access to them, will be at certain risk of heat stroke, collapse, and death. These will not be limited to those who need to work outside; those who can't afford air-conditioning, or to move to more northerly, cooler climates because of age, family relations, or lack of jobs, will all be at risk. Their numbers could be staggering, and our ability to respond effectively, severely limited and costly.

One relatively recent event suggests just how serious the health effects of warming can be. Chicago experienced an unprecedented weeklong heat wave in 1995. More than 700 people are thought to have died as a consequence. Hospitals were forced to shut down their emergency rooms, and the federal government dispatched refrigerator trucks to simply store the dead bodies, because the morgues had run out of room. This was one major and, at the time, unusual episode. We can anticipate many such instances in our future, affecting large swaths of the United States on a regular basis. Estimates suggest that by the end of the century one-third of our population will experience at least one 95 degree wet-bulb day every year, and at some locations of course, several such days, on average. There are obviously many more public health challenges that climate change will bring; I've just mentioned them for lack of statistically precise estimates. But they include, at a minimum,

- Storm surges with resulting flooding
- A literal "sinking" of the west coast and southern U.S., as rising sea levels sweep inland in many coastal cities around the world and cause major damage and displacement
- The potential dislocation of our traditional systems of food production and distribution, and therefore nutrition (though the ever-resourceful "agri-business" is likely to adapt)
- A dramatic increase in arthropod-borne diseases (dengue, West Nile fever, etc.) and perhaps even traditional viral and bacterial spread

The potential health impact on low- and middle-income countries can only be imagined. Bangladesh suffered a devastating cyclone disaster in 1970, causing the loss of nearly a quarter of a million people in a single night. Bangladesh is expected to become the poster child for the impact of rising water levels associated with climate change: a densely populated, infrastructure-poor country, with a height gradient northward from the Bay of Bengal of barely 1 inch per mile. Low- and middle-income countries around the world have the least ability to adapt and will suffer the most. Those in tropical areas will suffer the additional burden of the shifting ecology of arthropod and other infectious diseases, particularly malaria and quite likely, cholera, among others.

In summary, if we do not make dramatic changes in greenhouse emissions, the trajectory of climate change is reasonably predictable, in ways that are not good for the public's health.

**Dr. Alfred Sommer**
*Johns Hopkins Bloomberg School of Public Health*

# *Preface*

This book had humble beginnings as a thesis for Cesar Marolla's master's degree in sustainability and environmental management at Harvard University. I had the pleasure of being Cesar's thesis director, and I knew upon first reading that this was an important piece of work. *Climate Health Risks in Megacities: Sustainable Management and Strategic Planning* is the first book that puts forth a solution to pressing climate change issues that affect the world's most vulnerable population—the urban poor—in a way that avoids making climate change a partisan issue. Further, it makes use of International Organization for Standardization (ISO) frameworks as part of the solution, providing an objective, clear, and auditable process, with a defined process for continuous improvement.

Climate change is a particularly difficult issue to address. First, there is the sheer scope of it. While weather at the local level may be consistent, hotter, or colder, it is affected by change happening on a global level. Second, those who caused a majority of the emissions that led to this situation are not willing to pay a proportional share of the cost to mitigate the climate changes caused by their generational actions. Last, the uncertainty of what will actually happen leads many country leaders to pause before investing, since the timing of catastrophic climatic impacts is uncertain. Solving large, complex problems is often left for future generations. Those of us in the business of climate change feel that "kicking the can down the road" for future generations to address is not a viable solution.

The time to act is now, and several forward-thinking city mayors, spurred into action by witnessing the effects of climate change on their cities firsthand, have decided to act.

Hurricanes Katrina and Sandy showed that even major cities in the wealthiest nation in the world lack continuity plans, redundant systems, backup power, and communications. A simple storm can render a city like New York inoperable for weeks, if not months. Further, while there was broad scale effects felt everywhere, the effects were felt hardest by the working poor. There are stories even today of lower income people being displaced over a year after Sandy.

This book provides case studies in several megacities: New York, Los Angeles, Beijing, Rio de Janeiro, London, Mumbai, and Lagos. Cesar looks at key issues like physical location, proper sanitation, food security, and vector-borne diseases against the backdrop of climate change, and models its effect on the urban poor. For example, urban poor often inhabit the least desirable land, flood zones that are especially susceptible to climate events. Further, extreme rainfall variability affects crop production, which impacts food prices, which impacts the poor the most. Frequent flooding also tends to cause an increase in insect population, which causes an increase in vector-borne diseases like Chagas and malaria.

Addressing the issues and challenges associated with mitigating the effects of climate change is enormously difficult. However, this book lays out a framework for action. By using established international standards that provide step-by-step guidance, along with a compelling business case for action, cities that face harmful climate impacts can begin to systematically address these risks. Since this award-winning thesis has been published, Cesar has gotten the mayors of New York, Los Angeles, and Rio interested in this approach. By presenting an approach to address climate impact that avoids partisanship and instead focuses on developing a compelling business case, cities across the world can begin to lay out a long-term plan for action.

Standards-based solutions refer to the "plan–do–check–adjust" model for carrying out change. Many have planned climate mitigation and adaptation strategies, but often faced opposition if they did not address the issue comprehensively. The beauty of this book, and the ideas Cesar puts forward, is that it provides a plan—a starting point. However, this starting point provides many useful checks, balances, adjustments for constraints and, most important, an opportunity to adjust the approach. No plan is perfect, but the only safe plan is one never started. Opportunities to mitigate the effects of climate change are disappearing rapidly. The time for planning is over. The time to do is now. This book shows us how.

**Richard Goode**
*Executive Director, Ernst & Young, LLC*
*Adjunct Professor, Harvard University and Tufts University's Gordon Institute*

# *Acknowledgments*

Completing a book that deals with the magnitude and scale of the impact of climate change on the urban population health of megacities and presenting an all-inclusive risk management strategy to adapt and manage these effects is truly an ambitious task. The essence and beauty of this book is that although strategies are driven by numbers and extensive analysis, they are untainted by bias, judgment, or opinion.

I would not have been able to complete this work without the aid and support of countless people over the past 5 years. I must first express gratitude to my faculty members at Harvard University, whose leadership and expertise tackling global issues is exemplary. Furthermore, throughout this stimulating journey, I have had the pleasure of knowing a vast and diverse pool of people in the scientific community. I have interexchanged my music career and the scientific world for the past 20 years; traveling extensively around the world playing my music with major artists and for our American troops. It has been such an enriching experience of self-discovery. Moreover, exploring cities and researching, interviewing world leaders, and assessing cities' climate risks and vulnerabilities on urban dwellers have been highly rewarding, and this intersection of art and science represents a gold mine of creativity. Much to my surprise and delight, I found the link between art and science, where life experiences are intertwined and changed through our personal creativity and curiosity, deeply stimulating. The works of Roald Hoffmann, a theoretical chemist who won the 1981 Nobel Prize in Chemistry and who has also published plays and poetry, is personally inspiring. The world needs more intellectual encouragement and creative minds. Leaders must display confidence and interest in the present and future generations' creative ideas by challenging the status quo. Nobel Laureate Dr. Walter Gilbert said: "One of the great dangers, one of the great problems, is, as fields grow, they stultify. They get drowned in the knowledge." As presented in this book, strategic and analytical thinking and flexibility have to be relentlessly applied and even challenged in seeking new viewpoints because this is the best way to find the direction and agility to adapt and cope with new challenges.

Furthermore, I am deeply humbled and grateful for the insights of our military leaders. The Department of Defense understands and is taking leading actions to identify the most serious and climate-related security risks for each Combatant Command. The Defense Department is looking at the ways in which the Combatant Commands are integrating mitigation of risks into their planning processes, and describing the resources required for an effective response. The work of the Honorable Katherine Hammack, assistant secretary of the Army for Installations, Energy and Environment, and her ability to provide the U.S. Army with clear direction and follow through with purpose is a force of transformation. Her leadership is exemplary!

Collaboration among several groups and individuals is often needed to address complex issues such as climate change, health risks, and urban populations' vulnerabilities. This effort was made with no financial support from any institution or individual. Therefore, I am deeply humbled and extremely honored to have had the help and support of the following leaders who have added highly valuable information to my work with their expertise:

**Judy Baker** is a lead economist in the Global Practice for Social, Urban, Rural and Resilience at the World Bank focusing on issues of urban development in East Asia and the Pacific. She has worked across regions, particularly in Latin America and South Asia in the World Bank, covering urbanization and its impacts, service delivery, urban poverty, policies, and programs for upgrading informal settlements, climate change, and impact evaluation. She has published five books and has authored numerous papers and country studies related to poverty and service delivery issues. She has also taught as a guest lecturer at a number of universities.

**Dr. Martha M. L. Barata** graduated in actuarial science and in economics science, holds a PhD in energy and environmental planning from the Institute of Postgraduate Studies and Research in Engineering, Federal University of Rio de Janeiro. She is a professor and researcher at Oswaldo Cruz Institute (IOC), at the National School of Public Health, which belongs to the Health Research Foundation (Fiocruz) of the Federal Health Ministry. She is a member of the Urban Climate Change Research Network (UCCRN), the steering group UCCRN of the Second Assessment Report on Climate Change and Cities, which was published in 2015, and lead author of the health chapter of the First and the Second UCCRN Assessment Report on Climate Change and Cities. She was also the lead author of chapters in the *Global Environmental Outlook 5* published by UNEP, 2012.

**Dr. Aaron Bernstein** is associate director of the Center for Health and the Global Environment at Harvard Medical School and lecturer all over the world on the importance of biodiversity to human health, from the EPA in China, to the U.S. National Institutes of Health, to the Library of Alexandria in Egypt.

**Dr. Wenyuan Chang** is State Key Laboratory of Atmospheric Boundary Layer Physics and Atmospheric Chemistry (LAPC), Institute of Atmospheric Sciences, Chinese Academy of Sciences, Beijing, China.

**Alan Cohn** is climate program director at the New York City Department of Environmental Protection, where he develops cost-effective strategies to advance resiliency and prioritize investments in water resource management. He leads efforts on flood protection; coordinates national and international climate resiliency initiatives; promotes green approaches to drainage and water quality improvement; and advances studies of climate change impacts on water supply, storm water management, and wastewater treatment. Cohn managed the development of the NYC Wastewater Resiliency Plan and contributed to the New York City Comprehensive Waterfront Plan, Green Infrastructure Plan, and PlaNYC: A Stronger, More Resilient New York. Cohn previously worked at the U.S. Environmental Protection Agency (EPA) in Washington, DC, where he coordinated development of the U.S. Climate Change Science Program report on sea-level rise impacts to the Mid-Atlantic coast and contributed to the endangerment finding that provides the basis for EPA regulation of greenhouse gases. Cohn received a bachelor of science in atmospheric science from Cornell University in 2004, and a master of science in atmospheric and oceanic science from the University of Maryland in 2006.

**Dr. Vincenzo Costigliola** has more than 40 years of medical practice after graduating in medicine at the University of Naples (Italy), and a postdegree specialization in anesthesiology and intensive care from the Università di Pisa (Italy). He is a founder and president of the European Medical Association (Belgium). He also cofounded the European Association for Preventive, Predictive and Personalized Medicine, and is its current president (Belgium). He was formerly the chief of medical services in the Italian Navy. Over the years, Costigliola has widened his specializations into rheumatology, dermatology, proctology, oncology, surgery, drug abuse, emergency treatment, and disaster action. Costigliola has also developed expertise in hospital organization, medical teaching methodology, and computer and audiovisual for the medical profession. He is the medical adviser to OTAN and W.E.U. Bruxelles. He is a member of the International Advisory Board of the King Abdulaziz University, Jeddah (Saudi Arabia); a board member of the European Biotechnology Thematic Network Association (Italy); and a member of the editorial advisory board of the *Journal of Psychiatry in Primary Care*. He has been a renowned speaker at numerous international conferences and congresses.

**Nathalie Crutzen**, who holds a PhD in business and economics, is an associate professor in Accenture Chair in Sustainable Strategy at HEC-Management School of the University of Liege. She is responsible for several academic activities (scientific research, academic courses, supervision of master's and PhD dissertations) in the fields of business strategy, performance management, and sustainability. She is also the academic coordinator of a multidisciplinary platform called "Liege Sustainability Management Platform." This platform brings together professors and researchers from various areas in management (human resource management, marketing, social entrepreneurship, supply chain management, etc.) as well as from other disciplines (economy, geography, architecture, environmental sciences, etc.) at the University of Liege. Crutzen has an international profile. She has a lot of contacts with other universities, business schools, and various organizations (businesses, cities) all over the world, especially concerning issues related to sustainability and strategic management. She has presented dozens of scientific papers at international conferences in different worldwide countries, and she has also published several articles in international scientific journals (*Journal of Cleaner Production, Review of Business and Economics, Revue Internationale PME,* and *Humanisme et Entreprise*) as well as several book chapters. Her main current research projects deal with the relationship between sustainability and strategic management, entrepreneurship, and innovation, as well as with the management of sustainable and smart cities.

**Ariel Durosky, BA**, is a posttraumatic stress disorder (PTSD) team research coordinator. She earned a BA in psychology at Emmanuel College. She previously conducted research related to sex differences in leadership styles. She has close family members who served in the military and is passionate about working with veterans.

**Dr. Kurt J. Engemann** is the director of the Center for Business Continuity and Risk Management and professor of information systems in the Hagan School of Business at Iona College. He has consulted professionally over the past 30 years in the area risk management decision modeling for major organizations and has been instrumental in the development and implementation of comprehensive business continuity management programs. Engemann is a Certified Business Continuity Professional (CBCP) with the Disaster Recovery Institute International. Engemann is the editor in chief of the *International Journal of Business Continuity and Risk Management* and the *International Journal of Technology, Policy and Management.* He teaches courses in the areas of business continuity, risk management, systems development, operations management, and decision support systems. Engemann is Distinguished Professor at The International Institute for Advanced Studies in Systems Research and Cybernetics. He has a PhD in operations research from New York University and has published extensively in the area of risk management and decision modeling.

**Dr. Julio Frenk**, a noted leader in global health and a renowned scholar, became the sixth president of the University of Miami on August 16, 2015. He also holds an academic appointment there as professor of public health sciences at the Leonard M. Miller School of Medicine. Prior to joining the University of Miami, Frenk was dean of the faculty at the Harvard T.H. Chan School of Public Health since January 2009. While at Harvard, he was also the T&G Angelopoulos Professor of Public Health and International Development, a joint appointment with the Harvard Kennedy School of Government. He served as the Minister of Health of Mexico from 2000 to 2006. There he pursued an ambitious agenda to reform the nation's health system and introduced a program of comprehensive universal coverage, known as Seguro Popular, which expanded access to health care for more than 55 million uninsured Mexicans.

Frenk holds a medical degree from the National University of Mexico, as well as a master of public health and a joint PhD in medical care organization and in sociology from the University of Michigan. He has been awarded honorary doctorates from several institutions of higher learning. In September 2008, Frenk received the Clinton Global Citizen Award for changing "the way practitioners and policy makers across the world think about health."

**Dr. Wendi Goldsmith** is cofounder and director of the Center for Urban Watershed Renewal. She is a Yale-trained geologist with 25 years of consulting roles guiding communities and engineering teams to adopt measures that better address sustainability and climate change. For the past 5 years, she has engaged with European researchers funded in part through insurance industry support to develop pragmatic strategies to guide climate adaptation and resilience. A pioneer in green infrastructure, Goldsmith has supported local, state, and federal agencies to carry out water resources engineering projects accommodating or harnessing natural processes. She played a lead role in the planning, design, and program management of the $14 billion post-Katrina Hurricane Storm Damage Risk Reduction System, the first regional-scale climate-adapted infrastructure system in the United States, informed by climate forecasts and related dynamics. She coordinated the science, policy, and engineering disciplines involved in implementing engineering design criteria for the new resilient infrastructure system in Greater New Orleans under Army Corps contracts totaling $200 million in services.

Goldsmith is known for building consensus among diverse and often antagonistic stakeholder groups to help advance large public projects. Her work exemplifies solutions operating on a systems level, addressing risk and resilience across uncertain scenarios. She has led research and development programs for the Department of Defense developing methods for evaluating and optimizing renewable energy and efficient/resilient buildings, infrastructure, and site design. She is well regarded for her expertise in the practical application of climate science in the built environment, notably within river corridors and coastal landforms. She has written and presented extensively on resilient design for climate change in urban settings and frameworks for decision-making informed by community engagement, including book chapters, peer-reviewed articles, reports, trainings, and keynotes.

**Richard Goode** is an executive director in Ernst & Young's U.S. Climate Change and Sustainability Services practice with experience in implementing sustainability programs at large organizations as well as expertise in carbon accounting. In this role, Goode assists in the development, verification, and implementation of climate change and sustainability assurance competencies throughout the United States. Previously he was the senior director of sustainability at Alcatel-Lucent. In this role Goode led the company's efforts in measuring, reporting, and setting $CO_2$ reduction targets, and creating and implementing sustainability programs across Alcatel-Lucent. Goode is an adjunct professor of sustainability at Harvard University and Tufts University's Gordon Institute. He is the founder of the Boston-area Sustainability Group and is a graduate of Presidio Graduate School in San Francisco, California, where he also serves on the board of directors. He also serves on the board of directors of the International Society of Sustainability Professionals (ISSP) and the board of advisors of Harvard University's Sustainable and Environmental Management program.

**Giovanni Grandoni** earned a PhD in physics at the University La Sapienza in Rome with a specialization in geophysics. Starting his career as a researcher in ENEA (Italian National Agency for New Technologies, Energy and Environment) in 1980, he has conducted investigations on meteorology and air quality. He has developed and implemented a multisources original numerical model about the pollutant dispersion and transport into the atmosphere and depositions to the ground level. His knowledge of relationships between air pollution and meteorology has significantly contributed to carry out a complex system of neural networks to predict and assess the air pollution levels by 72 hours in advance and to optimize effective actions of decision makers to prevent events of high pollution in urban areas.

As a result, with coauthors, he has patented A.T.M.O.S.FE.R.A. (Italian-language acronym of "analysis and treatment weather data with the aim of identifying statistically phenomena relating to air pollution"). Being a project and scientific leader of the Project A.T.M.O.S.FE.R.A., he has carried out activities and resources to develop neural network stochastic models to predict air quality in major Italian cities (Rome, Milan, Naples) where the hourly pollution level forecasting system is installed.

In the framework of the Convention UN/ECE on Long-Range Transboundary Air Pollution, he has designed and realized an updating information system to estimate ozone critical level maps for the vegetation (crops and forests) and the population in the overall Italian territory by analyzing the monitoring station network data.

In the ENEA Projects A.R.T.E.M.I.S.I.A. and A.R.T.E.M.I.S.I.A. 2, he has applied air quality models to finalize indicators for minimizing health and environment impacts from industrial plant airborne effluents in the context of the IPPC program (Integrated Pollution Prevention and Control).

As a project leader of air quality investigations over the shoreline industrial area of Biferno Valley (in Central Italy), he has coordinated the scientific activities and collaborated with Italian (Rome, Lecce) and foreign universities (Arizona State University [ASU], Finnish Meteorological Institute [FMI], Russian State Hydrometeorological University [RSHU], Helsinki) to characterize the micrometeorological state of the atmosphere into the planetary boundary layer. He also developed new technologies and scientific tools oriented to the air quality control.

He participated in project proposals (Call) of European Programs including LIFE+ and prepared proposals for postdoc positions, Marie Skłodowska-Curie actions, ENEA–2015. Moreover, he organized and participated in many experimental campaigns of meteorological measurements related to turbulence descriptions within the planetary boundary layer in urban areas and in open fields using remote sensing tools such as tethered balloons, SOnic Detection And Ranging (SODAR), Ceilometer with the MLH (Mixing Layer Height) module, etc.

Grandoni is a referee for international scientific journals specializing in meteorology and environment. He is the author and coauthor of numerous national and international publications, and has been also awarded for achieved outcomes by SAS Institute and by CIVR from the Department for Scientific Research.

**Kevin Robert Gurney, MS, MPP, PhD**, is an associate professor at Ecology, Evolution and Environmental Science School of Life Sciences, and a senior sustainability scientist, Global Institute of Sustainability, Arizona State University.

**Sir Andy Haines** is professor of public health and primary care, with a joint appointment in the Department of Social and Environmental Health Research and in the Department of Nutrition and Public Health Intervention Research, London. He was educated at Latymer

Upper School and at King's College London (MBBS, MD) where he qualified in medicine in 1969 with honors in pathology, surgery, and pharmacology and therapeutics. He earned an MD in epidemiology at the University of London in 1985. Under his leadership, the London School of Hygiene and Tropical Medicine received the 2009 Award for Global Health from the Bill and Melinda Gates Foundation worth $1 million for sustained commitment to improving the health of poor people, having been selected from 106 nominations worldwide by an international jury of experts. It was the first academic institution worldwide and the first UK institution to receive the award. Haines is a Fellow of the Royal College of General Practitioners, the Royal College of Physicians, and of the Faculty of Public Health, and was made a foreign associate member of the U.S. Institute of Medicine of the National Academies in 2007. He was knighted for services to medicine in 2005.

**The Honorable Katherine Hammack** was appointed the Assistant Secretary of the Army for Installations, Energy and Environment (ASA IE&E) by President Barack Obama on June 28, 2010. She is the primary adviser to the Secretary of the Army and Chief of Staff of the Army on all Army matters related to installation policy, oversight, and coordination of energy security and management. She is responsible for policy and oversight of sustainability and environmental initiatives; resource management, including design, military construction, operations, and maintenance; base realignment and closure (BRAC); privatization of Army family housing, lodging, real estate, and utilities; and the Army's installations safety and occupational health programs. Among her many accomplishments are the establishment of the Army's Net Zero program, and the creation of the Office of Energy Initiatives (OEI), which is working to streamline large-scale renewable energy projects to achieve 1GW of renewable energy by 2025.

**Anne Hilburn, BA,** is a research assistant currently working toward an MA in psychology at Columbia University Teachers College. She earned a BA at Hamilton College.

**Kevin W. Knight** is a member of the General Division of the Order of Australia and chair of the ISO working group that developed the new ISO 31000 risk management standard and the revision of ISO/IEC Guide 73, and a founding member of the Standards Australia/Standards New Zealand Joint Technical Committee OB/7 Risk management.

**Joe Leitmann, PhD,** is the World Bank's environment coordinator for Indonesia, with responsibility for activities related to climate change, natural resource management, pollution, global environmental issues, and environmental safeguards for all sectors. He is also the World Bank's disaster management coordinator and was the founder/first manager of the $650 million Multi Donor Fund for Aceh and Nias following the Indian Ocean tsunami. With over two decades at the World Bank, Leitmann has worked on projects, programs, and policies related to postdisaster reconstruction, natural resource management, urban environmental management, urban poverty, and renewable energy. In addition to short-term assignments in over 50 countries, he has had multiyear field assignments in Turkey, Brazil, and Indonesia. Leitmann began his career in development work as a U.S. Peace Corps volunteer. Leitmann holds a PhD in city and regional planning from the University of California/Berkeley and a master's degree in public policy from Harvard University's Kennedy School of Government. He is the author of *Sustaining Cities: Environmental Planning and Management in Urban Design* as well as numerous articles on sustainable development.

**Ari Lowell, PhD**, is a postdoctoral clinical researcher with a background in posttraumatic stress disorder (PTSD), depression, anxiety, and traumatic brain injury. He previously worked at the VA New Jersey Health Care System where he helped veterans suffering from the effects of combat-related trauma, military sexual trauma, and other difficulties. Lowell is experienced in both individual and group psychotherapy. He is a veteran of the Israel Defense Forces.

**Dr. Maria Cristina Mammarella** was born in Rome and graduated from La Sapienza University with a degree in mathematics. She holds a PhD in mathematical logic and astronomy. She has worked with Italian National Agency for New Technologies, Energy and Sustainable Economic Development (ENEA) since 1983 and its director of research since 2010. She specializes in weather studies concerning meteodiffusivity and air quality, and she made a significant contribution to work that determined, projected, and produced an automatic intelligent station based on neural network, able to forecast air pollution level 72 hours in advance. This effort created a new way to manage and control the air quality, and prevent critical events instead of being restricted to intervening after the occurrence. Owing to the originality of her idea, which combines the learning ability of neural network with the deterministic approach of weather forecasting, this intelligent station obtained the Italian certification with the name A.T.M.O.S.FE.R.A.©® Besides A.T.M.O.S.FE.R.A. projects, carried out in Rome, Milan, and Naples, Mammarella made, or coordinated, several ground-applied research projects for air quality control, such as A.R.T.E.M.I.S.I.A. in Udine district and A.R.T.E.M.I.S.I.A. 2 in Sicily. The results of this project are collected in the book *A.R.T.E.M.I.S.I.A. 2: Applicazione alla zona di Milazzo*, edited by M. C. Mammarella, with all researchers of the project as contributors and published by ENEA. She focused her research job on linking forecast systems to neural networks using the study of the "atmospheric boundary layer," which has fundamental effects on spreading pollution agents in the air, and to this purpose, she is in contact with Professor S. Zilitinkevich, of the Finnish Meteorological Institute (FMI) in Helsinki. In this thematic area, she is involved with international research groups, such as Arizona State University, the Notre Dame University (Indiana), the Finnish Meteorological Institute, and Helsinki University, for developing new instruments and technologies. In Italy, she is committed to the project for air quality in the industrial zone of Biferno Valley. She is a member of the Scientific Committee of the Strategic Plan for Sustainable Mobility City of Rome, and also 2015 Marie Skłodowska-Curie actions: post-doc positions in ENEA. From 2012 to 2013 she was the Italian representative to the European Community project TEMPUS for the development of meteorology and its new application. Russia and Eastern European countries collaborate with TEMPUS. She was a promoter of the workshop NATO ARW (Advanced Research Workshop) "Climate Change, Human Health and National Security," April 2011, Dubrovnik, Croatia. Credits for the results obtained were given to her by OCSE from Paris, by SAS Institute, and by CIVR from Department for Scientific Research. She was proposed as a candidate from "Accademia dei Lincei" for the Italgas Award and in 2008 she obtained the Eunomia Award. She is a referee for several international scientific journals specializing in meteorology and environment. Her research, besides being published in scientific journals and conference proceedings, receives attention from newspapers, and national and international reviews.

**Eric Maskin** is Adams University Professor at Harvard. He received the 2007 Nobel Memorial Prize in Economics (with L. Hurwicz and R. Myerson) for laying the foundations of mechanism design theory. He also has made contributions to game theory, contract theory, social choice theory, political economy, and other areas of economics. He earned an AB and a PhD at Harvard and was a postdoctoral fellow at Jesus College, Cambridge University. He was a faculty member at MIT from 1977 to 1984, Harvard from 1985 to 2000, and the Institute for Advanced Study from 2000 to 2011. He rejoined the Harvard faculty in 2012.

**Shagun Mehrotra** is assistant professor of environmental policy and sustainability management, The New School; and formerly managing director of Climate and Cities, an international policy advisory facility, CCSR, jointly housed at The Earth Institute, Columbia University, and NASA Goddard Institute for Space Studies.

**Dr. Holmes E. Miller** joined the department in 1991. He has a BS in industrial engineering and an MS and a PhD in management science, all from Northwestern University. His dissertation focused on personnel scheduling and allocation in health systems. He currently teaches courses in the areas of operations management, management science, electronic commerce, information systems, and sustainability. He also teaches courses in Muhlenberg's Accelerated Degree Program and has taught first-year seminars titled "Quality," "Learning, Working, and Knowing," "Money and Meaning," "Confronting Disaster," and "Why We Work," and is currently teaching "Business and Society" as part of the Business and Public Health cluster. Prior to coming to Muhlenberg, Miller spent 15 years in the business world, working for Chase Manhattan Bank and for the Union Carbide Corporation. He also was on the faculty of Rensselaer Polytechnic Institute. Prior to receiving his PhD, he worked in the steel industry, the railroad industry, and in management consulting. His current research interests include managing risk in manufacturing and service environments and in supply chains, both in terms of applying analytical methods to arrive at more informed and better decisions. He also is interested in sustainability as practiced in business environments, simulation modeling, and decision-making methodologies. He has published over 50 papers in journals, book chapters, and conference proceedings.

**Yuval Neria, MD,** is a professor of medical psychology in the Departments of Psychiatry and Epidemiology at Columbia University Medical Center and director of trauma and PTSD (posttraumatic stress disorder) at the New York State Psychiatric Institute. His research ranges from studying the mental health aftermath of high impact trauma (e.g., combat, terrorism, disasters), to translational studies aiming to identify neural and behavioral mechanisms of PTSD and neuroscience-informed treatment development. Neria grew up in Israel and personally experienced the effects of war. He is the recipient of the Medal of Valor, equivalent to a medal of honor. Neria has studied the effects of trauma across different contexts, including wars and disasters, and has been involved in developing and testing innovative treatments for PTSD. Neria's studies have been funded by the National Institute of Mental Health (NIMH), National Alliance for Research on Schizophrenia and Depression (NARSAD), and private foundations since 2002. Currently, his lab is using various neuroimaging methods (magnetic resonance imaging [MRI], functional MRI [fMRI], cerebral blood volume [CBV]) to identify brain markers of deficient fear conditioning, extinction, and overgeneralization aiming to probe highly needed brain and behavioral markers of trauma and PTSD. Neria has authored more than

140 articles and book chapters, and edited three textbooks published by Cambridge University Press. He is the editor in chief of the journal *Disaster Health.*

**Dr. Rajendra K. Pachauri** is the executive vice chairman of the Governing Council of New Delhi-based The Energy and Resources Institute (TERI), a position that he started on February 8, 2016. Before that, he was the chief executive of TERI beginning in 1982, first as director, and then director-general from April 2001 to February 7, 2016. Since 1998, he has also been chancellor, TERI University. In April 2002, he was elected as chairman of the Intergovernmental Panel on Climate Change (IPCC) and was reelected in September 2008. He continued in this position until February 2015. The IPCC, along with former U.S. Vice President Al Gore, was awarded the Nobel Peace Prize for the year 2007. He was senior adviser to Yale Climate and Energy Institute (YCEI) from July 2012 through June 2014; before that he was the founding director of YCEI (July 2009–June 2012). Pachauri has a PhD in industrial engineering and economics.

**Eduardo Paes** is a law school graduate of PUC-Rio University. He has dedicated his entire career to public service, starting in 1993 when he was appointed deputy mayor of the Jacarépaguá and Barra districts in Rio de Janeiro at the age of 23. In 1996, he was elected alderman and, in 1998, was elected to the Federal Chamber of Deputies. In 2000, he was appointed Rio's Municipal Environmental Secretary where he refined his knowledge of environmental issues. During his second term as a representative in Congress, in 2007, Paes was named Secretary of State for Tourism, Sport and Leisure, and helped in the organization of the Pan American Games that were held in Rio. A year later, Paes was elected mayor of Rio de Janeiro and on October 7, 2012, he was reelected in the first round for a second term.

Paes has led a comprehensive transformation of the city of Rio, addressing many structural challenges. His administration was able to achieve good results on education based on the permanent monitoring and evaluation of the students' learning progress. The program "Schools of Tomorrow" has shown consistent evolution, targeting stronger education in low-income areas of the city, among other initiatives. On health, the mayor has promoted a large expansion of primary care, increasing this coverage by more than seven times since the beginning of his administration, and currently includes 1.6 million people. Regarding mobility, he is promoting the largest expansion of mass transportation in the history of the city, with the construction of 152 km of Bus Rapid Transit lines as well as a complete reformulation in urban mobility, adding express corridors and bicycle lanes. Paes has focused on addressing long-term environmental issues. During his administration, the Gramacho landfill, one of the largest environmental liabilities in Rio's metro-politan area, was replaced by the Seropédica waste treatment center, helping to avoid pollution and carbon emissions. With sustainable urban development as a priority, the social housing program Morar Carioca benefits the low-income communities of Rio de Janeiro and won the Siemens Sustainable Community award in 2013.

Good management practices implemented in the Paes adminis-tration has earned Rio de Janeiro international recognition. The city, having earned investment grade rating from respected rating agen-cies, was identified as one of the best cities that responded to the international financial crisis and was the first in the world to enter into a credit transaction directly with the World Bank for economic and social development. As mayor, Paes was one of the leaders of the campaign that won Rio de Janeiro the right to host the 2016 Summer Olympics and Paralympics, the first Games to be held in South America. Today, in partnership with the state and federal govern-ments, he is working hard to ensure that major international events leave an important sustainable legacy for the city and improve the quality of life for all Cariocas (residents of Rio).

All that effort was recognized by C40 cities, and in 2013, Paes was unanimously elected as the C40 Climate Leadership Chair follow-ing the successful tenure of Mayor Michael Bloomberg of New York City. Paes received the Innovator Award Cycle Mayor 2009/2010. The initiative, sponsored by the Competitive Brazil Movement (MBC), Microsoft Brazil, Brazil and Symnetics Intel recognizes the applica-tion of good management practices in the municipal public service. The city of Rio recently won the World Smart City Award in 2013 at the Smart City Expo World Congress in Barcelona.

**Leon Edward Panetta** is cofounder of The Panetta Institute for Public Policy. He served as the United States 23rd Secretary of Defense from July 2011 to February 2013. Before joining the Department of Defense, Panetta served as the director of the Central Intelligence Agency from February 2009 to June 2011. Panetta led the agency and managed human intelligence and open source collection programs on behalf of the intelligence community. Panetta holds a bachelor of arts in political science and a law degree, both from Santa Clara University.

**Romel Pascual**, executive director, CicLAvia, has been a key fig-ure at CicLAvia from the start. As deputy mayor for Energy and Environment during the administration of former Los Angeles Mayor Antonio Villaraigosa, his leadership at the city helped the organization and the event gain traction within the city in CicLAvia's fledgling years. Pascual successfully merged the mayor's vision and CicLAvia's mission to help create what is now a city institution that transforms communities throughout the region. Before being named executive director in 2015, Pascual served on the CicLAvia board and was actively involved in the expansion of the organization.

Pascual has a master's in urban planning and has a long history championing environmental, energy, sustainability, and social justice issues in both the nonprofit and the public sector. He was the first assistant secretary of Environmental Justice under former California Governor Gray Davis. He led the environmental justice program at the U.S. Environmental Protection Agency Region 9–Western U.S., Hawaii and the Pacific Territories.

**Dr. Robert Pojasek,** president of Pojasek & Associates, is an internationally recognized expert on the topic of business sustainability and process improvement. He assists clients with developing and facilitating the planning and implementation of sustainability programs and sustainability management systems at both the corporate and facility level. Management systems help make sustainability programs a part of what every employee does every day. Pojasek has extensive experience with the implementation of a variety of management system standards including quality, environment, occupational health and safety, corporate social responsibility, and sustainability. He utilizes combinations of conventional management systems (ISO 9001, ISO 14001, and OHSAS 18001), risk management (ISO 31000), social responsibility (ISO 26000 and AS 8303), sustainability (BS 8900), business excellence frameworks (e.g., Baldrige Performance Excellence), and process improvement (Lean and Six Sigma). Pojasek has prepared business continuity and pollution prevention plans. He has also prepared corporate responsibility reports and applications for Dow Jones sustainability index recognition and is also experienced with implementing the U.S. Environmental Protection Agency's National Enforcement Investigation Center (NEIC) compliance-focused management system and the Occupational Safety and Health Administration Voluntary Protection Program (OSHA VPP).

With over 35 years of experience in the environmental, health, and safety consulting field, Pojasek has worked with a diverse range of clients in the manufacturing and service sectors and for nongovernmental organizations (NGOs) and government agencies. During this time, he has been very active in the practice of pollution prevention and has presented numerous conference presentations and written one hundred publications on pollution prevention and sustainability practices. He currently writes a blog on sustainability on the website greenbiz.com. As an adjunct professor at Harvard University, he teaches the popular distance-learning course "Strategies for Sustainability Management," and serves as a thesis director for students conducting research in sustainability at the master's degree level.

**Rodrigo Rosa** is the special advisor to the C40 chair and the executive representative of the City of Rio de Janeiro in the C40 Climate Leadership Group. He represents the chair in situations when Mayor Eduardo Paes cannot participate himself. Since 2009, Rosa has been serving as special advisor to the Mayor of Rio de Janeiro, Eduardo Paes, on sustainability and innovation policies. Currently he leads the strategic initiatives in the Office of the Mayor, which include the Rio Low Carbon City Development Program, a partnership between the City of Rio and World Bank, as well as climate change and resilience actions. Rosa also leads strategic international relations partnerships for the City of Rio. In 2010, Rosa was responsible in the municipality to organize the United Nations Sustainable Development Summit Rio+20 as well as the GRI Report (Global Reporting Initiative) among other activities. Rosa was also responsible for the development of the Rio Resilient initiative and was a collaborator of the Municipal Legislation on Climate Change, adopted in 2011 by City Council. Starting his career as a business journalist and economic analyst for major newspapers in Brazil, Rosa also worked as a legislative consultant in Congress prior to his work at Rio de Janeiro City Hall. He is currently a PhD student in environmental planning at Coppe University and holds a master's in economics of public sector and a bachelor degree in journalism from the University of Brasilia and a master business executive in environmental management at the University of Rio de Janeiro.

**Alfred Sommer, MD, MHS,** is a prominent professor of ophthalmology at the Wilmer Eye Institute and dean emeritus and professor of epidemiology and international health at the Johns Hopkins Bloomberg School of Public Health. He proved that a vitamin A deficiency dramatically increased childhood morbidity and mortality from infectious disease and that a 4-cent dose of vitamin A not only prevented and cured eye disease but also reduced childhood deaths by 34%. Because of his work, the World Health Organization, UNICEF, and their partners now annually provide more than 400 million vitamin A supplements to children around the world, saving hundreds of thousands of lives each year. Sommer is the recipient of numerous honors including the Albert Lasker Clinical Medical Research Award and the Spirit of Helen Keller Award, and is a member of both the National Academy of Sciences and the Institute of Medicine. He served as dean of the Bloomberg School from 1990 through 2005. Sommer received his MD from Harvard Medical School and his master of health science in epidemiology from the Johns Hopkins

School of Public Health. The Hopkins Sommer Scholars Program is named in his honor.

**Jean-Pascal van Ypersele** has a PhD in physics from the Belgian Université catholique de Louvain (UCL, 1986 with highest honors), based on work done at the U.S. National Center for Atmospheric Research (NCAR; Boulder, Colorado) on the effect of global warming on Antarctic sea ice. He has specialized in climate change modeling and the study of the impact of human activities on climate. He has authored papers on the modeling of sea ice; of paleoclimates; of the climate of the 20th and 21st century; and of regional climate in Europe, Greenland, and Africa. His more recent work is related to integrated assessment modeling of climate stabilization, and he works with economists in an interdisciplinary perspective. He has also published on the relations between climate and desertification, and the impacts of climate change on human activities and ecosystems. In 2008, he published a report on the mainstreaming of climate change adaptation and mitigation policies in the Belgian development cooperation. As professor at UCL (www.climate.be), he teaches climatology, climate modeling, mathematical geography, and environmental sciences, and directs the interdisciplinary master's program in science and management of the environment (www.uclouvain.be/cgse). He is the author of numerous scientific articles and popular works regarding climate change and sustainable development. He was a lead author for the Working Group II (WGII) contribution to the Third Assessment Report of the IPCC and was elected in 2002 vice chair of its Working Group II. For the IPCC Fourth Assessment, in addition to his responsibilities as member of the IPCC Bureau, he was also active in the Steering Group of the Task Group on New Emission Scenarios. He has participated in dozens of outreach events related to the IPCC work, in Belgium and abroad (including Brazil, Canada, France, Italy, Peru, Poland, Russia, Spain, the Netherlands, and United States), and is regularly interviewed by the media (Belgian and international) on climate, environment, and sustainable development issues.

**Robert W. Wood** retired from the U.S. civil service on March 1, 2013, after 42 years. He began his federal career in 1970 serving aboard the USS *Juneau* during the Vietnam War. After receiving an honorable discharge from the navy, he was hired by the Los Angeles District, Army Corps of Engineers. For the next 12 years he planned parks and flood control projects, wrote environmental impact statements, issued permits, and developed emergency and national defense contingency plans. In 1985, Wood accepted a position with the Plans and Programs Office at Edwards Air Force Base where he worked on long-range planning efforts. In 1986, he was asked by the commander to establish a new environmental organization. During the ensuing 27 years he expanded the civil service staff from 3 to 45 employees and hired over 86 support contractors who kept the base in complete compliance with over 400 regulated environmental activities in the three counties that encompass the installation and the six additional counties where test missions are flown. He shifted the organization from being compliance-driven to functioning as an environmental consultant providing real-time environmental compliance solutions to organizations across the base. By focusing on understanding of customer needs, his staff was successful in creating the conditions that allowed environmental requirements to be seamlessly integrated within routine workplace operations. Wood graduated in 1970 with a bachelor of arts in geography, University of Kansas, Lawrence, Kansas, and has completed graduate work at California State University, Long Beach, and earned a certificate in urban development, University of Southern California.

The following people helped to shape and complete the manuscript:

**Prudence Brighton** has more than 40 years of experience as a writer and editor, working first as a journalist and then as a technical communicator. She has worked for several well-known high technology companies, including Cisco Systems. Brighton has edited numerous master's degree theses for students in Harvard University's sustainability program. She holds a BA in history from Chatham University, an MS in management from Lesley University, and a graduate certificate in technical communication from University of Massachusetts Lowell.

**Steve Dunning**'s passion for helping others achieve their potential led to his work with many organizations in the veteran space. For the Department of Defense, he has shepherded groups through active theaters of operation across the Middle East. His lifelong respect and admiration for our nation's military only magnified with these first-hand experiences in the field with all branches of our active Armed Forces. He serves on multiple boards and steering committees such as the Los Angeles Mayor's Veterans Initiative, USC Veterans Collaborative, and Cal Poly Pomona Council to name a few. He is a frequent international and domestic speaker. During his 5 years internally at NBCUniversal, he was chosen to be the national colead of the company-wide diversity group for veterans and their allies. His goals with the NBCUniversal Veterans Network and the Got Your Six campaign, as an original steering committee member, aim to deeply impact our military, past, present, and future, and their families in ways most needed. Changing the national mindset of what it means to be a veteran and their incredible assets; they are trained, motivated, and highly qualified to lead, contribute, and uphold and grow our nation's great heritage. Dunning received the "Outstanding Citizen Award" selected and presented by the 38th Chief of Staff of the United States Army, Four-Star General Raymond T. Odierno. The award is presented to civilians not employed by the army, and cannot be presented to federal government officials at the policy-development level or technical personnel who assist the army as a consultant or adviser. It is the third-highest honor within the Department of the Army that may be presented to a private citizen.

**Christina Jonas Kennedy** is an author, researcher, and lecturer. Her research topics include business operations management, human resource management, and corporate social responsibility. She earned a master of science degree in administration with an emphasis in contracting and human resource management and her bachelor of science degree in business administration with an emphasis in management, both from California State University, Bakersfield. Also, she earned two associate degrees, one in business administration, and one in letters, arts, and sciences, both from Antelope Valley College.

# Author

**Cesar Marolla** is a sustainability and environmental management expert, researcher, author, adviser, speaker, and lecturer. Throughout his career he has brought strategy concepts to bear on many of the most challenging problems facing corporations and societies, including working with global organizations, such as Deutsche Telekom and nongovernmental organizations (NGOs) as well as volunteering with the U.S. Department of Defense. He has traveled extensively and worked in sustainability, climate change, and risk management, business marketing strategies, and corporate responsibility in Europe, South America, Middle East, Northeast Africa, and the United States.

Marolla has assembled sustainability assessments and best practices for Fortune 500 corporations and participated in climate talks with international organizations dedicated to communicating how the information and communication technology (ICT) industries can address and provide solutions for environmental issues. He has interviewed, researched, and collaborated with world-renowned leaders from the World Health Organization (WHO), Deutsche Telekom, U.S. Centers for Disease Control and Prevention (CDC), United Nations, World Bank, Columbia University, and Harvard University in addition to city mayors in issues such as climate change adaptation and mitigation strategies, environmental management, risk management, and public health risks.

Marolla is the recipient of the 2013 Harvard University Derek Bok Civic Prize Award that recognizes creative initiatives in community service and long-standing records of civic achievement. He also received a Certificate of Appreciation from the Pentagon for his volunteerism helping U.S. troops in the Middle East and Northeast Africa under the umbrella of the Global War on Terror (GWOT). Moreover, he received a "Military Coin," which is given as a token of affiliation, support, patronage, respect, honor, and gratitude, and presented by the Camp Victory Commander in Kuwait, Lieutenant Colonel Lawrence J. Smith.

Marolla earned a bachelor of science in business administration and marketing at Columbia College and a master's degree in sustainability and environmental management at Harvard University. He is also a graduate of the Executive Program in Sustainability Leadership at Harvard T.H. Chan School of Public Health—The Center for Health and the Global Environment.

# Section I

# Managing Climate Change and Health Risks in Megacities

# 1

## *Introduction*

A city is a system of systems. This model applies to the dependency of the city's resilience to its systems, such as transportation, utilities, and public health infrastructure and communication networks. These systems are individually dependent and also rely on how interdependent those systems are between them.

The stability of the global economy and political order are critically dependent on two inextricable factors: megacities and public health. Protecting the public health of these urban areas has more far-reaching repercussions than protecting the health of individuals in those densely populated areas. A health problem in one megacity can spread rapidly to other parts of the world (Khan and Pappas, 2011), leading to widespread public health concerns and destabilizing the megacity's operations. The effects are felt not just in the health sector but throughout the entire system.

As rapid urbanization is presenting new human and environmental outcomes, we are presently witnessing that cities are the future of humankind. Therefore, city governments must lead decisive roles in adaptation and mitigation strategies to climate change. Arguably, they have the main responsibility to lead in adaptation within their jurisdictions with the support of the regulatory, institutional, and financial structure from higher levels of government to efficiently deter and adapt to climate change impacts. The question then arises as to how well prepared city leaders are to adapt and minimize climate change effects on their populations' health. Most city leaders understand that their cities face the possibility of disruptive events from climate change and climate variability impacts that varies from brief interruptions of operations to the destruction of their infrastructure, the proliferation of diseases, and the loss of lives. Cities' operational frameworks supporting a strategy that includes prevention, preparedness, response, and recovery (such as risk management) strategies are implemented with a short-lived view of importance to the environments and the risks associated with those environments. Risk management and continuity management strategies to climate change adaptation remains a plan at the discretion of city leaders where lack of accountability, inefficient implementation, and disunited authority are common factors for failure.

Cities are taking an unnecessary gamble with their futures. Regardless of population density, modern or obsolete infrastructure, geographical location, and level of vulnerability to climatic events, cities need to identify their critical business components and effectively manage risks. Disaster recovery plans should not only identify and include redundant systems, backup sites, and alternative work sites. It is highly important to have in place a framework to operate the city back to "normal conditions" that includes processes for maintaining specific levels of communication, operation efficiency and safety from the time disaster strikes until business returns to normal.

Urban environments have implications—both positive and negative—for almost every phase of human health and well-being. Megacities can offer a wide range of public health services and innovations, in addition to the networks necessary to inform and educate the public on disease prevention, risks, and treatments. These benefits can drastically improve the quality of life in urban areas. However, megacities are also areas of deteriorated physical and social environments. These megacities frequently have high levels of air pollution, high building densities, and fewer social connections, which all have negative impacts on the health of urban residents (Harpham, 2009; Stephens, Satterthwaite, and Kris, 2008). This deterioration is especially common in the megacities of developing countries, where rapid urbanization has growing numbers of people living in slums and unhealthy environments. Furthermore, these risks will intensify as climate change exacerbates the vulnerabilities of the health and well-being of urban residents.

In planning for a comprehensive strategy of resilience to the devastating impacts of climate change in cities, we can draw from history. Along with the obvious damage to public and private structures and large numbers of people endangered, Hurricanes Katrina and Sandy in the United States and Typhoon Haiyan in the Philippines show the intensity of exposure to natural disasters, determining the effects of climate change on health, which are likely to be predominately negative. The number of health risks can be devastating. As in the case of Typhoon Haiyan, key issues immediately following the disaster were respiratory infections associated with overcrowding, especially acute respiratory infections in children. Measles were also a potential problem as well as vector-borne diseases, especially dengue and chikungunya fevers. Malaria was more endemic in the southern area and the country problems associated with childbirth and pregnancy in addition to malnutrition, especially of infants, were present (World Health Organization, 2013b).

Natural and human-caused disasters can leave a devastating impact on people's mental health. Psychological harm is closely correlated to personal injuries. A person going through the experience of the death of a family member, friend, or neighbor because of the disaster can be a strong forecast of a mental health condition influx. Researchers have pointed out displacement, migration, and the loss of personal assets and financial stability as factors that contribute to mental health issues, such as posttraumatic stress disorder (PTSD) and depression.

Environmental issues caused by severe climatic events are still unaccounted for in the economic system and not fully recognized as a high-priority risk in the health sector in most countries. Cities, and in particular megacities because of their population density, accelerated urbanization, and large infrastructure network, need to understand this new reality and invest in preparedness and risk management. The main challenge I witnessed throughout my research is the absence of a strategy that amalgamates all the risks associated with climate variability and change. The lack of a concise risk management approach to climate change reflects a clear signal: business continuity planning is not a high priority. Cities need to develop an international framework that materializes risk as a main concern. Presently, risk analysis models analyze risks independently and are static. Therefore, existing frameworks that operate in major cities don't account for the effect of one risk on another and offer no projections on how the risks might act in the future.

Mitigation strategies to deter climate change effects have received more international attention to date than adaptation to climate change and risk management strategies. Obviously, greenhouse gas emissions must be reduced to slow the effects of warming and climate change but, as we face an uncertain future, managing risks becomes vital for our continued existence. Risk management strategies should consider not only ways to lessen risks, but how cities should best adapt to the far-reaching effects of climate change impacts on health. Risk management provides cities with a universal framework and help map the path toward meaningful actions to combat the biggest global-health threat of the 21st century.

We must inspire a genuine sense of urgency, and, through a network of change leaders, empower others to act. I truly believe that creativity, curiosity, courage, and convicted are the essence of a leader. We must stand in the power of purpose to "Inspire, Lead, and Empower Change."

## List of Terms

**Climate change:** Change in the state of the climate that can be identified (e.g., using statistical tests) by changes in the mean and/or the variability of its properties, and that persists for an extended period, typically decades or longer. It refers to any change in climate over time, whether due to natural variability or as a result of human activity.

**Megacity:** A metropolitan area with a total population of more than 10 million people. The definition was expanded by Spanish sociologist Manuel Castells in response to economic and social shifts. Castells defines megacities as large urban areas in which some people are connected to global information flows while others are disconnected and information poor.

**Urban:** A nonagricultural production base with a minimum population size (often 5,000). Exact definitions vary from country to country (United Nations Statistics Division). Urban areas include continuously built-up inner city areas, as well as transitional or peri-urban areas between fully built-up and predominantly agriculture use or rural areas.

# 2

## The Intergovernmental Panel on Climate Change (IPCC) Fifth Assessment Report and Its Implications for Human Health and Urban Areas

R. K. Pachauri

The Fifth Assessment Report (AR5) of the Intergovernmental Panel on Climate Change (IPCC) has substantially increased our understanding of the scientific aspects of climate change. Of particular relevance to the discussion in this chapter is the assessment that the AR5 provided on the impacts of climate change as they relate to specific sectors of the economy, and how, through adequate action involving adaptation as well and mitigation, risks to human beings and other species can be minimized. The AR5 found that human influence on the climate system is clear and that anthropogenic emissions of greenhouse gases are the highest in history. Recent climate changes have had widespread impacts on human and natural systems (IPCC, 2014a).

The chief cause of human-induced climate change is an increased concentration of greenhouse gases (GHGs) since the advent of the industrial age. According to NASA's report *The Relentless Rise of Carbon Dioxide* (2013), ancient air bubbles trapped in ice allow researchers to go back in time and identify the condition of the earth's atmosphere and climate in the distant past. This analysis of carbon dioxide ($CO_2$) allowed scientists to determine that the gas was now present in higher concentrations than at any time in the past 400,000 years. Since NASA published its report, anthropogenic GHG emissions have increased, driven primarily by economic and population growth, and are today higher than ever. Atmospheric concentrations of carbon dioxide, methane, and nitrous oxide now are levels unprecedented in at least the last 800,000 years (IPCC, 2014c). Scientists have detected the effects of these gases, and other human-caused emissions, throughout the climate system. These emissions are extremely likely to have been the dominant cause of the observed warming since the mid-20th century.

It is evident that climate change can expose many parts of the world to extreme events, which, quite apart from the increase in warming and its associated impacts, are likely to expose humans to greater risks of morbidity and mortality. The AR5 predicted a rise in surface temperature during the 21st century under each emission scenario that it assessed (IPCC, 2014b). Heat waves are likely to occur more often and last longer, and extreme precipitation events will become more intense and frequent in many regions. The ocean will continue to warm and acidify, and the global mean sea level will continue to rise. Climate change will amplify existing risks and create new ones for natural and human systems. Disadvantaged people and communities will feel the effects disproportionately.

From 1901 to 2010, the global mean sea level rose by 0.19 meters, and the rate of sea-level rise since the mid-19th century has been larger than the mean rate during the previous two millennia. Each of the last three decades has been successively warmer at the earth's surface than any preceding decade since 1850. The period from 1983 to 2012 was likely the warmest 30-year interval of the last 1,400 years in the Northern Hemisphere, where such assessment is possible. The globally averaged combined land and ocean surface temperature data, as calculated by a linear trend, show a warming of 0.85°C from 1880 to 2012 (IPCC, 2014b).

Projections through 2050 indicate that climate change will affect human health principally by worsening the conditions that already exist. Throughout this century, health problems in many regions are likely to rise, especially in developing countries with low incomes, as compared to a baseline without climate change.

In the AR5, the IPCC presented four specific scenarios, or representative concentration pathways (RCPs). The highest emissions scenario, designated as RCP 8.5, calls for no particular intervention to mitigate GHG emissions, thereby leading to the largest GHG concentrations. In this case, the combination of high temperature and humidity in some areas for parts of the year is expected to compromise ordinary human activities, including growing food and working outdoors. In urban areas, climate change is projected to increase risks for people, assets, economies, and ecosystems, including hazards from heat stress, storms and extreme precipitation, inland and coastal flooding, landslides, air pollution, drought, water scarcity, sea-level rise, and storm surges. For those who lack essential infrastructure and services or live in exposed areas, these risks are amplified (IPCC, 2014a).

Many aspects of climate change, and any associated impacts, will continue for centuries, even if the anthropogenic emissions of GHGs cease. The risks of abrupt or irreversible changes increase as the magnitude of the warming increases. Warming will persist beyond 2100 under all four scenarios except the lowest concentration scenario, designated as RCP 2.6. For example, it is virtually certain that the global mean sea-level rise will continue for many centuries, with the amount of rise dependent on future emissions. The threshold for the loss of the Greenland ice sheets over a millennium or more, and an associated sea-level rise of up to 7 meters, is greater than about 1°C but less than 4°C of global warming with respect to preindustrial temperatures (IPCC, 2014a). Magnitudes and rates of climate change associated with medium to high emission scenarios pose an increased risk of abrupt and irreversible regional-scale shift in the composition, structure, and function of marine, terrestrial, and freshwater ecosystems, including wetlands. A reduction in the extent of permafrost is virtually certain with continued rise in global temperatures.

Adaptation and mitigation are complementary strategies for reducing and managing the risks of climate change. Substantial emissions reductions over the next few decades can cut climate risks in this century and beyond, increase prospects for effective adaptation, ease the costs and challenges of mitigation in the longer term, and contribute to climate-resilient pathways for sustainable development. Making effective decisions to limit climate change and its effects can be informed by a wide range of analytical approaches for evaluating expected risks and benefits, recognizing the importance of governance, ethical dimensions, equity, value judgments, economic assessments, and diverse perceptions and responses to risk and uncertainty.

Risks to the climate result from how various hazards (including hazardous events and trends) interact with the vulnerability and exposure of human and natural systems, including their ability to adapt. Rising rates and magnitudes of warming and other changes in the climate system, accompanied by ocean acidification, increase the risk of severe, pervasive, and in some cases irreversible detrimental impacts. Some risks are specific to individual regions, whereas others are global. It is possible to reduce the overall risk of future impacts by limiting the rate and magnitude of climate change, including ocean acidification. The precise levels of climate change sufficient to trigger abrupt and irreversible change remain uncertain, but the risk associated with crossing such thresholds increases with rising temperature.

Many species face increased risk of extinction due to climate change during this century and centuries to come. This is especially true as climate change interacts with other stressors. Most plant species cannot naturally shift their geographical ranges sufficiently fast to keep up with current and high projected rates of climate change in most landscapes. Future risk is indicated to be high by the observation that natural global climate change at rates lower than current anthropogenic climate change caused significant ecosystem shifts and species extinctions during the past millions of years. Marine organisms will face progressively lower oxygen levels and high rates and magnitudes of ocean acidification, with associated risks exacerbated by more water temperature extremes. Coral reefs and polar ecosystems are highly vulnerable. Coastal systems and low-lying areas are at risk from sea-level rise, which will continue for centuries even if the global mean temperature is stabilized (IPCC, 2014a). The mega-deltas, such as Shanghai, Dhaka, and Kolkata, are particularly vulnerable, with very large and dense population characteristics in each city.

Climate change is anticipated to cause an increased displacement of people. Populations that lack the resources for planned migration experience higher exposure to extreme weather events, particularly in developing countries with low income. Climate change can indirectly increase risks of violent conflicts by amplifying well-documented drivers of these conflicts such as poverty and economic shocks.

Adaptation can mitigate the impact of climate change, but there are limits to its effectiveness, especially with greater magnitudes and rates of climate change. In the context of sustainable development, taking a long view now increases the chances that more immediate adaptive actions will improve future options and preparedness.

Adaptation can contribute to the well-being of a population, the security of assets, and the maintenance of ecosystem goods, functions, and services now and in the future. Adaptation is place and context specific. A first step toward adaptation is reducing our vulnerability and exposure to present climate variability.

Integration of adaptation into planning, including policy design, and decision-making can promote synergies with development and disaster risk reduction. Building adaptive capacity is crucial for effective selection and implementation of adaptation options. Complementary actions at all levels of society, from individuals to governments, can enhance adaptation planning and implementation. National governments can coordinate adaptation efforts of local and subnational governments, for example, by protecting vulnerable groups; by supporting economic diversification; and by providing information, policy, and legal frameworks, and financial support. Local government and the private sector are increasingly recognized as critical to progress, given their roles in scaling up adaptation of communities, households, and civil society and in managing risk information and financing.

Without additional mitigation efforts beyond those in place today, and even with adaptation, the risk of severe, widespread and irreversible global effects by the end of the 21st century is high to very high. Mitigation involves the possibility of adverse side effects, but these hazards would not be as severe, increasing the benefits from near-term mitigation efforts.

Delaying additional relief to 2030 would substantially increase the challenges associated with limiting warming over this century to below 2°C relative to preindustrial levels. It would require even higher reductions in emissions from 2030 to 2050; a much more rapid scale-up of low-carbon energy over this period; a larger reliance on carbon dioxide cuts in the long term; and greater economic impacts in the medium and long term. Estimated global emissions levels in 2020 based on the Cancun Pledges are not consistent with cost-effective mitigation trajectories that are at least about as likely as not to limit warming to below 2°C relative to preindustrial levels, but they do not preclude the option to meet this goal (IPCC, 2014a).

In principle, mechanisms that set a carbon price, including cap and trade systems and carbon taxes, can achieve mitigation in a cost-effective way, but have been implemented with diverse effects due in part to national circumstances as well as policy design. In the short term, loose caps have limited the impact of cap and trade systems. In some countries, tax-based policies specifically aimed at reducing GHG emissions—alongside technologies and other policies—have helped to weaken the link between GHG emissions and gross domestic product (GDP). In addition, in a large group of countries, fuel taxes have had effects that are akin to sectoral carbon taxes.

Behavior, lifestyle, and culture have a considerable influence on energy use and associated emissions, with high mitigation potential in some sectors, in particular when complementing technological and structural change. It is possible to significantly lower emissions by changing consumption patterns, adopting energy saving measures, changing diets, and reducing food waste.

We can pursue strategies and actions now that will move toward climate-resilient pathways for sustainable development, while at the same time helping to improve livelihoods, social and economic well-being, and effective environmental management. In some cases, economic diversification can be an important element of such strategies. The effectiveness of integrated responses can be enhanced by relevant tools, appropriate governance structures, and adequate institutional and human capacity. Integrated responses are especially relevant to energy planning and implementation; interactions among water, food, energy, and biological carbon sequestration; and urban planning, which provides substantial opportunities for enhanced resilience, reduced emissions, and more sustainable development.

With the predicted effects of climate change becoming more serious, urban areas, which intrinsically have greater institutional strengths and infrastructure for attaining climate resilience than other locations, must focus on responses to current and future climate change for the well-being and healthy living of citizens on Planet Earth.

Climate change is a threat to sustainable development. Nonetheless, there are many opportunities to link mitigation, adaptation, and the pursuit of other societal objectives through integrated responses. Successful implementation relies on relevant tools, suitable governance structures, and enhanced capacity to respond.

Most important, successful implementation would require surmounting inertia of various kinds—scientific, informational, institutional, and behavioral. As the Fourth Assessment Report (AR4) of the IPCC indicated, the transformation of energy supply also must contend with vested interests. Dealing with the challenge of climate change would require actions not confined to technological solutions, changes in economic decision-making, or transformation of institutions. Significant shifts would require changes in attitudes, value systems, and lifestyles. Even though the AR5 concluded that behavior, lifestyle, and culture have a "considerable" influence on energy use and associated emissions, these would require substantiation through the assessment of various options and pathways available to human society. The world has to evaluate responses to climate change based on conditions today and those projected for the future. Solutions need to be devised in a comprehensive manner to include actions that have thus far received inadequate attention, but go to the heart of human behavior, values, and ethics.

## List of Terms

**Acidification:** The significant changes to the chemistry of the ocean that occurs when carbon dioxide gas (or $CO_2$) is absorbed by the ocean and reacts with seawater to produce acid.

**Anthropogenic:** Relating to or resulting from the influence of human beings on nature.

**Greenhouse gas (GHG):** An atmospheric gas, either manmade or natural, that contributes to the greenhouse effect by absorbing thermal infrared radiation emitted by the earth's surface, the atmosphere, and clouds. The main greenhouse gases are water vapor, carbon dioxide, nitrous oxide, methane, and ozone.

**Gross domestic product (GDP):** The total value of goods produced and services provided in a country during one year.

**Intergovernmental Panel on Climate Change (IPCC):** The international body that assesses the science of climate change. The IPCC was set up in 1988 by the World Meteorological Organization (WMO) and United Nations Environment Programme (UNEP).

**Peri-urban:** The environs immediately adjoining an urban area; between the suburbs and the countryside.

**Representative concentration pathway (RCP):** Four greenhouse gas concentration trajectories adopted by the IPCC for its Fifth Assessment Report (AR5) in 2014, superseding the Report on Emissions Scenarios (SRES) projections published in 2000.

# 3

## The Urban Poor and the Response of Megacities to Dealing with Climate Change Health Risks

Megacities demonstrate an intricate response to the impacts of climate change on the health of the urban poor. Their reactions are due to the unpreparedness of the public and the authorities in dealing with the consequences. Therefore, the ability of a megacity to develop adaptation strategies is restrained by the inability of city leaders to make collective decisions, resulting in a lack of coordinated efforts and government inaction.

The predicament is clear to see and straightforward to recognize but hard to unravel. Despite the economic success of megacities, their leaders are still seeking solutions to the health consequences of climate change. They are trying to determine what the impacts are for the urban poor, how to respond with preparedness and adaptation strategies, and what collective actions are available to reduce the health risks associated with climate change.

Climate change is a major concern that is closely related to our good health. Today, it presents health risks even more profound than those posed by poor sanitation and living conditions in 19th-century London (Epstein and Ferber, 2011). A population's health depends on environmental stability and the functioning of the world's climate system. In the next decades, most cities will experience the impact of climate change, and major cities around the world will gradually face the consequences. The increase in risk factors will put the lives and safety of billions of people in jeopardy, but the threat will affect the population unequally. Climate change will compound existing vulnerabilities, increase poverty levels, and affect the urban poor, because they typically live in the most vulnerable areas of a city and are at high risk from the effects of changing climate and natural disasters. Overcrowded living conditions, inaccessibility to safe infrastructure, and poor health conditions make the urban poor highly susceptible to the impacts of climate change (Baker, 2012c).

A changing climate can alter the pattern of diseases, mortality, human settlements, food, water, and sanitation. Climate change brings increasing temperatures, rising seas, and more frequent incidence of severe storms. These effects produce dangerous sanitary events and are known health risks (Jensen, 2007). In addition, shifting rainfall patterns may influence transmission of many diseases, including water-related illnesses, such as diarrhea, and vector-borne infections, including malaria. Finally, climate change could have far-reaching effects on how food is produced and may have health impacts from increasing rates of malnutrition.

*The Anatomy of a Silent Crisis* (*Human Impact Report: Climate Change*) (Global Humanitarian Forum, 2009) confirmed that every year more than 300,000 people die of the consequences of climate change, 325 million people are seriously affected, and economic losses total US$125 billion. Four billion people are vulnerable, and 500 million people are at extreme risk. Higher temperatures such as the heat wave across Western Europe in 2003 resulted in more than 35,000 fatalities; the estimated economic loss of the 2011 Thailand floods was US$30 billion and of Hurricane Katrina US$125 billion; and the Horn of Africa droughts in 2011 claimed tens of thousands of lives and threatened the livelihoods of 9.5 million people. Moreover, the destructive Hurricane Sandy that landed near Atlantic City, New Jersey, on October 29, 2012, caused 113 deaths, left up to 40,000 people homeless with economic losses that are estimated at $70 billion dollars, with approximately 200,000 homes damaged. The exposure of cities to financial and environmental losses as a consequence of climatic events is evident, and leaders must embrace bold, coordinated international action to reduce the impacts of climate change.

The environmental effects of a changing climate include compromised water supplies, decreased air quality, diminished food security, and direct effects on vectors (Portier et al., 2010). As rainfall patterns shift, the severity and occurrence of droughts will increase in some regions while floods will increase in others. These events will impact the vector population as climate change may alter the distribution of vector species—increasing or decreasing the ranges, considering favorable or unfavorable weather conditions that alter the breeding places (e.g., vegetation, host, or water availability) (Intergovernmental Panel on Climate Change [IPCC] Working Group II, 2001). Thus, altered precipitation patterns could change the population density of vector-borne diseases, infecting inhabitants in greater numbers and affecting the health of entire communities. Flooding will affect roads, schools, power supplies and houses, destroying essential infrastructure that supports communities and cities (Portier et al., 2010). Heavy rains will force storm water discharge systems to empty waste into the water supply. Rising sea levels will affect cities near the coasts, which consequently will result in salination of freshwater supplies, loss of productive land, and a change in coastal dwelling mosquitoes' breeding (Epstein, 2001).

In Africa alone, by 2020, projections state that between 75 million and 250 million people will face an increase of water stress due to climate change. Water security problems are also expected to intensify by the year 2030 in southern and eastern Australia. In the course of the century, water supplies stored in glaciers and snow cover will likely decline, reducing water availability in regions supplied by melt-water from major mountain ranges (such as the Himalayas in Asia and the Andes in Latin America). In North America, decreased snowpack in western mountains is projected to cause more winter flooding and reduced summer flows, exacerbating competition for over-allocated water resources (Van Ypersele, 2007). Freshwater availability in Central, South, East, and Southeast Asia, particularly in large river basins, may decrease due to climate change, which could, in combination with other factors, have an adverse impact on more than a billion people by the 2050s (Van Ypersele, 2007). Furthermore, air pollutants are changed by higher temperatures. The higher temperatures also affect the range and concentration of pollen, affecting a larger portion of the population (Portier et al., 2010).

Ground-level ozone ($O_3$) and fine particulate matter (PM2.5) are associated with an increased risk of mortality. Although significant progress has been made in reducing ambient concentrations of air pollution in the United States, recent levels of $O_3$ and PM2.5 remain elevated from the natural background and are within the range of concentrations found by epidemiology studies to affect health (Fann et al., 2011). In general, the percentage of all deaths due to PM2.5 exposure is higher than it is for ozone. Southern California has the highest percentage of PM2.5-related deaths among the United States population. If the air quality incrementally improves, the number of deaths related to PM2.5 and ozone changes. For example, 23,000 PM2.5-related mortalities would be avoided by lowering 2005 annual mean PM2.5 levels to 10 $\mu g/m^3$ (micrograms per cubic meter of air) nationwide. The estimate concludes that approximately 80,000 premature deaths would be preventable by lowering PM2.5 levels to 5 $\mu g/m^3$ on a national scale (Fann et al., 2011).

Food yields will be affected by climate change. Though climate will affect populations around the world, it will have a greater impact primarily in regions where warmer temperatures and reduced rainfall occur. Coastal cities will be affected by a reduction in food supply caused by rising sea levels, heavy storms and floods, and a reduction of arable land. These negative factors are increased by acidification, warmer waters, and reduced river flows (Bates et al., 2008). A global increase in the distribution and incidence of infectious diseases is projected because of climate change effects, which will lead to crises with economic and social implications. Malaria and other vector-transmitted diseases get most of the attention from climate-change scientists and health professionals. The latitudinal, altitudinal, seasonal, and interannual connection between climate change and disease, along with historical and experimental evidence, leads to the conclusion that climate, in addition to other causes, significantly influences infectious diseases in a nonlinear fashion (Lafferty, 2009).

Climate change will alter the patterns and spread of malaria transmission. Rainfall affects malaria, acting not only on persistent bodies of water, but also on physical and biochemical characteristics of aquatic environments. Heavy rains and flooding are known to cause major malaria outbreaks in semiarid or arid lowlands; at the same time, spatial, and temporal variations in rainfall determine the nature and scale of malaria transmission in highland areas (Gilioli and Mariani, 2011). Furthermore, Chagas disease is one of the most significant climate-sensitive vector-borne diseases in South America, and it is spreading throughout the continent. The globalization of the disease through climate change as well as other factors, such as migration (it can be increased by climate change impacts in certain regions), blood transfusion, and so forth, is a concern for developed countries, and these aspects are modifying and accelerating the transmission rate and distribution of the disease.

The relation between Chagas disease and temperature change has been studied almost since the beginning of the disease's parasitological, clinical, and epidemiological description. Direct impacts such as thermal stress, floods, and storms; and direct impacts such as borne-vector diseases, water-borne pathogens, water quality, air quality, and food availability and quality are concerns for urban populations. The socioeconomic situations of the urban residents as well as the environmental conditions of the city are important factors that amplify the impacts of climate change on a population's health (World Health Organization [WHO], 2003). Another factor to assess the vulnerability of populations is the existing city's adaptation and mitigation strategies and the viability of the institutions, technology in place, and risk management planning to be implemented when the city's leaders need to act upon the situation (WHO, 2003).

Although little work has happened to lessen climate change impacts on public health and its propagation into the urban ecosystem, the awareness of risk management has been emphasized considerably in the government and the private sector. The devastating consequences for densely populated cities of large-scale disasters such as Hurricane Sandy, which had overwhelming impacts in October 2012 affecting 24 states, including the entire eastern seaboard from Florida to Maine, with particularly severe damage in New Jersey and New York, presents a challenge to public policy if meaningful actions are not taken into consideration. Climate change impacts on megacities is a "global shock" that allows for a unique outline of risks.

These types of global shocks stream a series of risks that grow to be active threats because they extend its effects across global systems. These threats expand along the urban ecosystem arising in health, environmental, social, or financial risks.

In pursuing these complex issues, research was conducted using the case study method to explore climate change health impacts on the urban poor in megacities. The research looked at practical and theoretical cases, differentiated by geographic, demographic, and socioeconomic factors. The case studies focused on the following megacities: New York, Beijing, Los Angeles, Rio de Janeiro, Mumbai, London, and Lagos. These cities have different economic and social driving forces, and they are presently implementing strategies to combat climate change with different approaches and results. The evidence shows that a risk assessment and management strategy, when well planned and modeled on the needs and conditions of each city, will provide a viable advantage in combatting climate change and its health impacts on the urban poor.

## Urbanization and the Environment

The proportion of people living in urban areas has risen to 50% and will continue to grow to two-thirds, or 6 billion people, by 2050. Most of the global gross domestic product (GDP) will be devoted to providing energy and resources to the new city residents over the next decades. As the cities' structures change to accommodate the new wave of dwellers, these factors will have a deep impact on the environment and human health (Burdett and Sudjic, 2011).

Distorted landscapes are common features of urban areas. Megacities are far from natural ecosystems and processes. This situation leads to altered microclimates in addition to other effects, such as considerably elevated surface and air temperatures (Blake et al., 2011). The urban features of the landscape obstruct long-wave radiative cooling processes. Because of the vertical geometry of the urban areas, the ventilation that circulates between buildings, houses, and construction sites is generally reduced (Blake et al., 2011).

Urbanization affects the environment to such an extent that it changes the microclimate; creates loss of agricultural land; contaminates air, soil and water; and increases water use and runoff (Rosenzweig et al., 2011a). Albedo is "the fraction of solar energy (shortwave radiation) reflected from the Earth back into space and it is a measure of the reflectivity of the earth's surface" (Institute for Global Environmental Strategies, 2009). Because cities have dark asphalt roadways, rooftops, and urban-canyon light trapping, a city's albedo is significantly lower than natural surfaces. Therefore, cities trap heat and do not release it back to the surface, creating a heat island effect. Furthermore, solid surfaces replace natural soil and vegetation, reducing evapotranspiration and latent heat cooling. This net cooling effect and an increase in ambient urban carbon dioxide levels create a detrimental environment that is unhealthy for all (Blake et al., 2011).

## The Impact of Climate Change on Megacities

The changing climate is already affecting cities and, according to projections, the effects will only get worse (Carbon Disclosure Project, 2011). Cities are motionless and this makes them vulnerable to the impacts of climate variability (World Bank, 2010b).

As the occurrence of extreme weather events increases year after year, the acute vulnerability of urban populations, especially children, the elderly, and the poor, will be exacerbated by climate change. The interruption of critical supplies to urban populations is a serious threat that is aggravated by the cities' dependency on food imports (World Bank, 2010a).

Air quality will adversely affect health and safety. Health risks among those who are heat-sensitive will increase. The vulnerabilities of the urban poor will be intensified by heat fatalities due to poor air circulation in congested slums and lack of access to air conditioning (Rodriguez et al., 2009).

C40 is a network of the world's megacities committed to addressing climate change. Practically every C40 network of megacities has determined they are exposed to physical risks that endanger its population. The C40 group has identified risks from climate change effects that are already happening or that are expected to take place in the near future. Most of the cities have identified several risks, such as more intense rainfall, increased severity of storms and floods, and rising sea levels. Furthermore, there are many compounding factors that will aggravate the physical effects of climate change in the city (Carbon Disclosure Project, 2011). Table 3.1 lists the global effects of climate change identified by the cities.

Megacities create drastic changes in biological and physical systems, as well as hazards affecting the health outcomes of the population. Climate change differs from other global hazards because it is spatially and temporally diffuse. It affects future generations and goes beyond any boundaries; accountability is also diffuse because it is difficult to hold one entity directly responsible (Rosenthal et al., 2007). Some serious issues require an immediate response from our world leaders. The World Bank has conducted client-cities studies for many years and has concluded that poverty is a more pressing and troublesome problem than climate change. However, climate change is the most complex issue society faces today and it has global connotations. The required capability for cities and world leaders to deal with climate change presents a big challenge (World Bank, 2010b).

The changing climate impacts most city residents, particularly the needy, in both rich and poor countries. Urbanization and climate change are inextricably connected because the consequences of the latter can detrimentally affect urban areas to the extent that entire systems are disrupted (World Bank, 2010b). Migration to major cities is increasing rapidly, and problems and opportunities will grow in both importance and complexity. How effectively cities respond to climate change will provide critical insight into how to adapt to other complex issues the world will face in coming decades due

**TABLE 3.1**

Physical Effects of Climate Change Identified by Cities

| Effect | % of Respondents | Top 3 Sectors Affected |
|---|---|---|
| Temperature increase/heat waves | 85% | Human health, energy, water |
| More frequent/intense rainfall | 79% | Buildings, water, transport |
| Sea-level rise | 67% | Buildings, waste, transport |
| Storms and floods | 58% | Human health, buildings, water |
| Drought | 42% | Water, human health, energy |

*Source:* Carbon Disclosure Project, 2011, *Global Report on C40 Cities.*

to urbanization (World Bank, 2010a). Some observers believe that poverty is on an accelerated path, particularly in the developing world. As the distinguished scientific journalist and publisher Gerard Piel told an international conference in 1996: "The world's poor once huddled largely in rural areas. In the modern world they have gravitated to the cities." The rising numbers of the poor living in urban areas is known as "the urbanization of poverty" (Piel, 1997).

The rapid increase of the poverty-stricken in megacities is a particular characteristic of urban expansion. Also, in many developing countries the population of the urban poor is increasing in higher proportions than the overall population growth. This disparity results in more inequalities and economic and social injustice (Vlahov et al., 2011). The urban poor will be the most affected, because the cost of staple foods increases as climate change creates droughts, heavy rains, and affects agricultural production diminishing supplies. The socioeconomic groups affected by climate change will likely see an increase in poverty rates that are parallel to extreme weather events (Gardner et al., 2009). The effects of urbanization are more or less leading to the scarcity of health services, but primarily to maldevelopment (Vlahov et al., 2011). Maldevelopment is a qualitative notion that refers to a discrepancy between the needs of a specific population and the responses generated to meet those needs (Frenk et al., 2010).

The poor are particularly affected by climate change, and the health impacts on the population are predominantly experienced in major cities around the world. Megacities are already besieged by the growing population and the services they need to provide to their residents. Taking into consideration that megacities need to accommodate more than 3 million new habitants every week, the resources, prevention, and adaptation measures are exhausted already without more residents overwhelming major cities' basic services every day (Abeygunawardena et al., 2003).

The number of poor people living in urban areas has been rising in all regions of the world with the exception of Europe and Central Asia (ECA).

## Natural Disasters and Poverty

Historical records show that megacities developed in rich countries. Currently, however, most of the world's megacities are in relatively poor countries. Tokyo, New York, Los Angeles, Paris, Osaka, and London are the megacities in high-income countries and, in the 20th century, most of the largest urban agglomerations were within the most advanced economies (Jedwab and Vollrath, 2014).

Megacities' populations keep growing at an accelerated pace. Even though the steady march of urbanization brought global economic prosperity to some, many megacities remain "poor." One of the reasons is another characteristic of today's megacities: They have grown through natural increase (excess of births over deaths) instead of in-migration, which was the contributor to growth previously. As a result, increasing populations have put downward pressure on wages. Thus, the megacity remains poor and its size has no correlation to its productivity (Jedwab and Vollrath, 2014).

According to the World Bank report *Shock Waves: Managing the Impacts of Climate Change on Poverty* (Hallegatte et al., 2016), poor households are more exposed to floods than other households. The poor settle in risky areas due to lower prices and land scarcity. The correlation between poverty and natural disasters detrimentally affects city dwellers because of exposure and vulnerability, as natural disasters increase inequality and may contribute to a decoupling of economic growth and poverty reduction. Hence, disasters are found to aggravate poverty (Hallegatte et al., 2016). The Harvard University Center for International Development conducted a study titled *Impact of Natural Disasters on Human Development and Poverty at the Municipal Level in Mexico* (Rodriguez-Oreggia et al., 2010) and stated that poverty worsened in Mexico municipalities between the 2000 and 2005 floods and droughts. Food poverty increased by 3.6%, which is about an 8% increase in the food poverty of those municipalities who experienced a disaster. This figure represents a 5.8% increase on average for affected municipalities and asset poverty by 1.5%, which is about a 2.3% increase in poverty rates for areas impacted by the climatic event. Therefore, the study shows that natural disasters reduce human development and increase poverty, and this effect can be sizeable: The average impact on human development in the affected areas is similar to going back 2 years in terms of their human development gains over the 5-year period reviewed. Hence, the impact of natural disasters in the studied areas is higher with lower levels of human development, whereas the wealthier municipalities have not experienced the aforementioned (Rodriguez-Oreggia et al., 2010). Another example is Ethiopia's famine incidence in the 1980s. It took a decade on average for asset-poor households to bring livestock holdings back to prefamine levels after the 1984–1985 famine (Dercon, 2003). These events have shown the susceptibility of the urban poor to external factors that magnify the risks associated with poverty and climatic impacts.

## Urban Environmental Challenges and Health Repercussions

As stated earlier, megacities are not planned well enough to assimilate the accelerated growth in population. For example, water supply and waste disposal are becoming difficult to manage, because the existing infrastructure is decades old and cannot cope with the increasing demand (Nair, 2009). Untreated domestic wastes are a serious concern in the megacities of developing countries, because they cause surface and groundwater resources to deteriorate. Increasing industrialization and urbanization have not translated into effective practices for the dumping of solid waste, particularly in landfills, which remains the principal disposal and implied treatment method (Hamer, 2003). As social and ecological factors are reviewed for their impacts on health, it is urgent that the health sector examines its potential to deter the spread of infectious diseases, so that the world's health systems can respond adequately. Analyzing and identifying the impacts of climate change on health in urban areas is imperative. Urban populations, and especially the poor, will be affected differently due to the way factors such as age, location, housing infrastructure, health-care access, and preexisting health conditions determine outcomes. Consequently, poor sanitation and lack of basic services must be addressed to better prepare megacities for the impacts of a changing climate and to strengthen adaptation strategies for resilience. Environmental conditions and infectious diseases are fundamental factors to understand and differentiate among the wide range of ecological dimensions to infectious disease. Therefore, initiatives to reduce exposure to diseases play a critical role in people's survival and lessens the health impacts of climate change.

As health shocks play an important role in the dynamics of poverty, three core factors emerge:

- The main diseases that affect poor people are diseases that are expected to expand with climate change (such as malaria and diarrhea).
- Health expenditures are regressive, with poor households largely uninsured—such outlays push an estimated 100 million people per year into poverty—and the loss of income for the sick or the caregiver can have a large impact on family prospects (WHO, 2013a).
- Children are most vulnerable to these shocks and can suffer from irreversible impacts that affect their lifetime earnings and lead to the intergenerational transmission of poverty (Hallegatte et al., 2016).

Poverty, natural disasters, and public health are intertwined, and their synergies have a countereffect that threatens sustainable development. With challenges to the urban poor worsened by the impacts of climate change and natural hazards, strategies are required that will reduce risks by building their capabilities to cope with recurring disasters, strengthening communities to build capacities of urban governance, and ensuring urban dwellers can live on safe lands and have access to basic services, particularly after a disaster.

## List of Terms

**Albedo:** The fraction of solar energy (shortwave radiation) reflected from the earth back into space. It is a measure of the reflectivity of the earth's surface.

**Fine particulate matter (PM2.5):** Particle pollution (also called particulate matter or PM) is a mixture of solid particles and liquid droplets found in the air. Particle pollution includes "inhalable coarse particles," with diameters larger than 2.5 micrometers and smaller than 10 micrometers; and "fine particles," with diameters that are 2.5 micrometers and smaller.

**Ground-level ozone ($O_3$):** Ozone is found in two regions of the earth's atmosphere: at ground level and in the upper regions of the atmosphere. They have the same chemical composition ($O_3$). Ground-level ozone is not emitted directly into the air but is created by chemical reactions between oxides of nitrogen and volatile organic compounds.

**Maldevelopment:** An imbalance between the needs of a specific population and the responses generated to meet those needs.

**Risk assessment:** A methodology to determine the nature and extent of risk by analyzing potential hazards and evaluating conditions of vulnerability that together might harm exposed people, property, services, livelihoods, and the environment on which they depend.

**Salination:** Any process that increases the salt content of the ocean.

# 4

## Managing Climate Change and Health Risk in Megacities

Martha Barata

### Introduction

Megacities have grown fast in the last 20 years; many of them are in developing countries, and their high population density and dependence on a complex infrastructure makes them particularly vulnerable to the impacts of climate change. These transformations suggest a need to rethink the way they operate and adapt to their evolving environments.

A large and continuously expanding body of research shows that the way in which cities are planned and managed can make a substantial difference to the health of their residents (Rydin et al., 2012). Therefore, a public health system should advocate for policies, plans, and projects for all urban sectors. Rydin et al. (2012) goes on to say that "long-term projections of global health outcomes now explicitly include factors such as unsafe water, poor sanitation, urban air pollution, and indoor air pollution"—all of which are aggravated by climate change. In this context, health systems should be prepared for potentially enhanced disease risks related to climate change, which is expected to affect the frequency and severity of existing diseases (Smith et al., 2014).

This chapter focuses on how to use climate science and socioecological research to understand the potential impacts of climate hazards to urban residents and to plan strategies that can protect and enhance the adaptive and mitigative capacities of cities and their citizens. This chapter presents an overview of urban health outcomes and their climate-related drivers, strategies to manage climate health risk in megacities, and the challenges of the development of megacities without considering the climate health risk.

## Climate Hazards for Health Systems in Megacities

Climate change can affect the health of the population directly (physical effects on the human body) or indirectly. In the latter case, the weather affects the environment and society (economy, infrastructure, productive and recreational activities, housing, and health systems), resulting in side effects on health (Confalonieri and Margonari de Souza, 2015; Intergovernmental Panel on Climate Change [IPCC], 2007).

In megacities, the overall burden of illness and the deleterious effects of climate change are likely to be substantially magnified, due primarily to its high population densities. The characteristics of megacities, such as impermeability of hard surfaces, urban heat islands, increased pressure on local ecosystems for home building and infrastructure, concentration of wastes, air pollution, and the existence of underserviced urban slums, requires assessment and redress to overcome dwellers' vulnerability to climate change (Barata, Ligeti et al., 2011).

Storms, floods, heat extremes, and landslides are among the most frequent and significant weather-related disasters that can harm human health. They can lead to an increase of

- Deaths
- Endemic infection and gastrointestinal diseases, caused by their contribution to the creation of breeding sites for disease vectors and bacterial contamination
- Water-borne and food-borne disease, through contamination with chemical toxins
- Trauma, occupational accidents, and mental health problems

Other climate-related health problems can also occur in megacities, such as

- Nutritional deficiencies in poor areas where food security is affected by the weather
- Coughing, wheezing, and asthma, caused by moisture damage, indoor mold, and aeroallergens related to the onset of the spring pollen season in temperate zones (Rosenzweig et al., 2015)

Megacities are often the destination for people displaced from their rural livelihoods when climate variation leads to crop failures, with substantial potential for social disruption and poverty.

Poverty and poor health status are highly correlated. Growing economic inequality—whether as a result of climate change or other socioeconomic stressors such as rapid urbanization—will likely magnify the gulf between the health status of the wealthy and the poor (Satterthwaite et al., 2008;

Smith et al., 2014; Walters and Gaillard, 2014). Poor people in megacities commonly have limited access to healthy housing and high-quality health care (Satterthwaite et al., 2008). The everyday deficiencies in the infrastructure of informal settlements such as slums, meanwhile, can exacerbate the health impacts of extreme climate events in these highly populated areas. Considering climate change, the outbreaks of water-borne infectious diseases are one health concern of particular importance in slum areas, where crowding and poor sanitation are prevalent (Freitas et al., 2014; WHO, 2012).

Extreme climate events can lead to disruptions in critical infrastructure operations and affect the health of megacity dwellers. For example, the lack of electricity can make it difficult or impossible to control the interior ambient climate, refrigerate food, physically move about in high-rise buildings, pump water to upper floors, and operate medical support equipment (Beatty et al., 2006).

These infrastructure disruptions can lead to a wide range of adverse health effects depending on the age, health status, and economic resources of the residents in the affected households. For example, exposure to ambient heat or cold in the absence of climate control may lead to temperature-related illness or aggravate underlying chronic conditions.

During an extreme heat event, morbidity and mortality rates may be especially severe if a power outage (blackout) occurs. While a lack of air-conditioning in homes increases the risk of heat-related death (O'Neill et al., 2005; Wheeler et al., 2013), air-conditioning also contributes to higher power demand during heat waves, which increases the risk of power disruptions and blackouts. When blackouts occur, exposure to heat increases, with a corresponding increase in health risks.

Blackouts can also increase risk of carbon monoxide poisoning from improper use of generators and cooking equipment. During August 2003, for example, the largest blackout in U.S. history occurred in the Northeast. Although this particular blackout did not coincide with a heat wave, it occurred during warm weather and resulted in approximately 90 excess deaths and an increase in respiratory hospitalizations (Anderson and Bell, 2012; Lin et al., 2011). These are some examples of indirect health impacts caused by extreme climate events in megacities around the world.

## Strategies to Manage Climate Change Health Risk in Megacities

Climate change is challenging the mission of megacities' public health systems to promote physical and mental health, and prevent disease, injury, and disability (Ebi and Semenza, 2008). Managerial responsibility for climate health risk in cities, in the long term, is essential to achieving health for residents in a sustainable way.

City leaders need to understand the situation looking forward: the expected status of the municipality regarding its opportunities and anxieties in both internal and external environments. After that, they need to apply the best available tools and resources to plan and implement strategies and programs so that they can achieve medium- and long-term goals. For that, they should answer three questions:

- Where are we?
- Where do we want to go?
- How do we get there?

In this context, city managers committed to reducing the potential impact of climate change should engage strategic stakeholders to respond to three fundamental questions:

- How do we identify climate change hazards and their locations? Hazards can be identified using global and regional scenarios developed by climate scientists; most of them are members of the International Panel of Climate Change (IPCC). It is noteworthy that climate scientists still face uncertainty, even though this is a known science and fairly investigated (Oppenheimer et al., 2007). Many efforts are being developed to reduce those uncertainties in order to support city policies, plans, and programs.

- How do we mitigate those hazards? To mitigate hazards, cities should identify, plan, implement, and monitor strategies that can reduce the emission of greenhouse gas. Unfortunately, the reduction of emissions in one city does not mean that the effects of climate change will be reduced by the same proportion. Managers should always look to adopt mitigation strategies that take into account co-benefits for the health of city dwellers, for example, transportation strategies.

- How do we prepare the city to prevent or reduce the impacts considering those hazards? The term *adaptation* refers to the process of designing, implementing, monitoring, and evaluating strategies, policies, and measures intended to reduce climate change impacts and to take advantage of opportunities (Ebi and Semenza, 2008). In public health, the analogous term is *prevention*. Adaptation generally entails understanding a system vulnerability, and to plan and manage systems in response to an anticipated change (Moser and Ekstrom, 2010). Vulnerability has many different definitions. The most common, according to Adger (2006), is that a vulnerability consists of exposure and sensitivity to disturbances or external tensions and the ability to adapt.

Adaptation strategies adopted by different sectors may reduce impacts to the health system. It is noteworthy that effectively responding to climate change is a process, not a one-time assessment of risks and likely effective interventions (Ebi and Semenza, 2008). It is necessary to periodically assess the adequacy of the results on the way to achieving the goals set.

The Urban Climate Change Research Network (UCCRN) developed a vulnerability and risk management paradigm that can be a useful framework for city decision-makers to analyze how their cities should adapt to the anticipated impacts of climate change. Its implementation requires the knowledge of the hazards, the vulnerability, and the specific adaptive capacity for each city (Mehrotra et al., 2011). The vulnerability of urban residents is connected to a city's social, economic, environment or physical attributes, and adaptive capacity.

The UCCRN was established in 2007 and now has nearly 600 world-class researchers who are developing a global resource for generating and sharing knowledge about mitigation and adaptation to reduce climate risk in low-, middle-, and high-income cities. Their aim is to provide climate change information and advice to decision-makers for the cities in which they are based. Their expertise spans a wide range of specialties, including climate scientists, urban heat island and air quality experts, impact scientists, social scientists (including political scientists and economists), and urban planners and designers. Their research should provide knowledge for decision-makers to use to enhance policies based on climate science. As a result, there is growing research and policy interest in urban climate risk assessment.

A health impact assessment of policies, plans, and projects is critical when implementing adaptation strategies. Campbell-Lendrum and Corvalán (2007) suggest that promotion and adoption of "Healthy Cities" campaigns, for example, are simultaneously beneficial for climate mitigation, adaptation, and population health. Some general adaptation measures that could reduce the vulnerability of city dwellers to climate change listed in Barata, Ligeti et al. (2011) include

- Incorporating climate change information and projections into city policies, plans, and codes. Examples include those plans governing where and how to construct and update transportation and buildings infrastructure or to relocation of a population from floodplains and steep hillsides to safer locations to reduce deaths.

- Reducing impermeable surfaces by providing green space, urban trees, vegetation, and other means such as permeable pavements and green roof, and solar energy on the top of the buildings, reducing flooding and heat island in megacities.

- Extending and maintaining essential infrastructure services such as drinking water supply and sanitation, taking into account climate conditions; energy and water conservation programs that reduce pressure on these systems in conditions of extreme heat and drought, and assure the continuity of these services. Preparing those services to respond well in extreme event cases.

- Using adaptation strategies that present a co-benefit for the health of city residents. New York City offers some examples of such strategies applied by different sectors that can reduce the pressure on its health system. They are

  - "Developing Climate-Resilient Water and Wastewater Infrastructure in New York City" by Alan Cohn, a case study presented in Chapter 15 of this volume, represents a strategy implemented by an infrastructure sector of a megacity that has co-benefits to health and reduces dwellers' vulnerability to climate change.

  - The Million Trees NYC project. The goal of the project was to have in 2016, 1 million trees planted in Manhattan. It has almost been achieved. Some tree-related health benefits include averting heat-related morbidity and mortality by reducing ambient temperature, decreases in respiratory illness due to improvements in air quality, and a general increase in physical activity associated with green space. This project, however, could have a negative consequence by increasing allergic responses to tree pollen. Research is currently underway to determine the relationship between tree pollen prevalence and the incidence and prevalence of allergic diseases in New York City. As in the case of the benefits of trees, initial studies have determined that both the density of tree canopy and the specific species of tree are important factors to consider in assessing the allergenic potential of urban tree pollen (Kinney et al., 2014, as cited in ARC3.2; Lovasi et al., 2013).

- Understanding the considerable disparities in the health impact among megacities in different regions, and between them in the same regions, are expected. These disparities exist due to temperature swings, precipitation, storminess, and other characteristics of climate change, which will vary significantly among regions with diverse geographical, weather patterns, and demographics characteristics. They are also due to social, economic, environmental, and institutional conditions in each megacity (WHO, 2009, as cited in Barata, Ligeti et al., 2011).

- Identifying how vulnerable a given population is to climate change, namely, its propensity or predisposition to adverse effects, and knowing the key determinants of their vulnerability. The location of the most vulnerable population has immense practical significance for best planning adaptation strategies and prioritizing resource investment (IPCC, 2014b). Indicators that are summarized in an index are a relevant tool for planners and managers. They are increasingly being utilized in vulnerability assessments (Adger, 2006) to guide the development of adaptation policies and strategies (Eriksen and Kelly, 2007). A climate vulnerability index is typically derived by combining, with or without weighting, several indicators assumed to represent vulnerability (IPCC, 2012). It can be represented in maps and will be a useful tool to effectively aid the adoption of effective adaptation strategies, considering that the causes of vulnerability are well understood and included in the index (Eriksen and Kelly, 2007).

Therefore, mapping the multidimensional aspects of city dwellers' vulnerability to climate change is a useful tool to manage the health risks of climate change. It shows the most vulnerable population considering its exposure, sensitivity, and adaptive capacity; and it supports the adoption of adaptation strategies to reduce health system vulnerability. Confalonieri et al. (2009) initiated mapping population vulnerability to climate change for states in Brazil and implemented it in the northeast region of the country. Barata et al. developed similar work for municipalities of the State of Rio de Janeiro in Brazil (Barata, 2013; Barata, Confalonieri et al., 2011). Assessing city dwellers' vulnerability to climate hazards as well as the location of the most vulnerable population has immense practical significance for best planning adaptation strategies and prioritizing resource investment.

## Challenges for the Development of Megacities without Considering Climate Risks

One important consideration throughout is that megacities are complex systems. They have many distinct components with a wide range of potential interactions; and these components self-organize over time, exhibiting behavioral patterns that emerge from these diverse and complex interactions (Batty, 2008). Recognizing and exploring this complexity can clarify what management strategies may be most effective and may highlight knowledge gaps to be filled through modeling, targeted learning, and other strategies.

Some important general recommendations related to the impacts of climate change on health and the facilitation of the adaptation process to these changes are (adapted from Confalonieri et al., 2009)

- Apply and adapt the methodology used to map multidimensional aspects of dwellers' vulnerability to climate change.
- Develop an integrated information system for morbidity and mortality resulting from extreme weather events in the city. Such a system

would allow easy identification of victims of landslides and floods, their specific problems, immediate causes, and consequences.

- Improve control systems for epidemics and their vectors (mosquitoes and rodents), especially those sensitive to weather variations.
- Clarify public opinion about climate change and its risks to the health of the population of the megacity based on dependable studies and reliable models in order to avoid the common misinformation that takes place in this area.
- Engage strategic stakeholders in the process of managing the health risks of climate change in megacities.
- Install systems directed toward environmental, epidemiological, and entomological surveillance in selected areas and situations, for the purpose of early detection of signs of biological effects of climate change (e.g., in vector populations).
- Stimulate scientific study and technical evaluations on the local level that integrates the health sector with other sectors (e.g., housing, urbanization, demographics, climatology, air quality) to build urban scenarios for the coming decades.

## List of Terms

**Capacity:** The combination of all the strengths, attributes, and resources available within a community, society, or organization to achieve agreed goals.

**Hazard:** Natural hazard events can be characterized by their magnitude or intensity, speed of onset, duration, and area of extent.

**Risk management:** The systematic practice of managing uncertainty to minimize potential harm and loss. Risk management comprises risk assessment and analysis, and the implementation of strategies and actions to control, reduce, and transfer risks.

# 5

## Greenhouse Gas (GHG) Emissions and Climate Change Risks

Climate change and urbanization are two of the most important issues facing the world today. Associated with them are challenges that world leaders must confront: poverty reduction and sustainable development. These must remain the center of society's global priorities. According to the International Energy Agency (IEA), urban areas now account for more than 71% of energy-related global greenhouse gases (GHGs). This percentage, which is linked to the growth of urbanization and population density, will increase to 76% by 2030. Thus, energy-related emissions are the largest single source of GHGs when looking at allocated allowances for the areas in question (Hoornweg et al., 2011).

Urban residents and their associated possessions are likely to account for more than 80% of the world's total GHG emissions. Therefore, cities are blamed for the majority of GHG emissions. However, recent studies have demonstrated that populations living in denser city centers produce half the amount of GHGs compared to the suburban residents. The disparities among the largest cities also must be acknowledged because the emissions rates in developing countries are lower than in developed countries, and are particularly low among the poorest residents. As a comparative statistic within the cities analyzed in this research, Table 5.1 shows city-based GHG emissions per capita as reported by the Intergovernmental Panel on Climate Change (IPCC). Variations in these values derive from production- and consumption-based values for cities (Hoornweg et al., 2011).

Over the coming decades, a substantial economic incentive to mitigate global GHG emissions could offset the projected growth of GHGs or reduce them below current levels. All sectors and regions have the potential to contribute to the reductions. The largest potential is in the building sector. Some of the commercially available options assessed by IPCC are efficient lighting and day-lighting; more efficient electrical appliances and heating and cooling devices; improved cook stoves; improved insulation; passive and active solar design for heating and cooling; alternative refrigeration fluids; and recovery and recycle of fluorinated gases. The IPCC estimates that by 2030, about 30% of the projected GHG emissions in the building sector can be avoided with net economic benefit. Moreover, there are significant co-benefits, for example, improved indoor and outdoor air quality as well as advances in social welfare (Van Ypersele, 2007).

**TABLE 5.1**

GHG Baselines for Cities under Study

| City | Emissions (tCO$_2$e/capita) | Year |
|------|------------------------------|------|
| London (City of London) | 6.2 | 2006[c] |
| London (Greater London) | 9.6 | 2003[b] |
| Beijing | 10.1 | 2006[a] |
| Los Angeles | 13.1 | 2000[b] |
| New York City | 10.5 | 2005[b] |
| Rio de Janeiro | 2.1 | 1998[a] |

*Source:* Hoornweg, D. et al., 2011, Cities and greenhouse gas emissions: Moving forward, *International Institute for Environment and Development*, 23(1), 207–213.

*Note:* Value includes emissions from aviation and marine sources.

[a] (Kennedy et al., 2009a).
[b] (Kennedy et al., 2009b).
[c] (Mayor of London, 2007).

## A Model for Change

Cities and their leaders can serve as a model for change. The promotion of innovative ideas, mitigation plans for climate change, environmental efficiency, risk management, and frameworks that contribute to the well-being of the urban population will enhance an already established management plan and will enable prosperity through economic development (World Bank, 2010b).

Cities have the opportunity to combat climate change with local changes that have a global impact. That translates into a tangible approach to acting quickly to climate change events between the residents and cities' leaders. The development of a set of common adaptive measures, with a focus on climate-related natural disasters, creates the potential for big cities to consider a strategic plan that brings the strategy under a universal framework. Big cities have perhaps the most to lose, and, therefore, the political momentum to develop an emergency plan because their assets are at a high risk (Pascual, 2012).

## Climate Leadership

The network of C40 cities is working to minimize their carbon footprints and implement programs to reduce GHG emissions. Because these cities emit 300 million tons of carbon dioxide (CO$_2$) per year, the transport sector is critical to the carbon reduction program of the C40 group. Seventeen of the 36 responding cities have already implemented a comprehensive plan to reduce their $CO_2$ emissions. The relation between population density and levels of per-capita carbon emissions from transport indicates that some urban areas have a more harmful impact on the environment than others. Higher population density can translate into lower transport emissions and the correlation between GDP (gross domestic product) and per-capita emissions from transport in C40 cities is also acknowledged (ARUP/C40 Cities Climate Leadership Group, 2011).

Although population density, major cities, and transport emissions clearly are related, the statistics acquired in studies show many disparities among cities around the world. Some of the highest per capita GDPs are in the cities of London, Hong Kong, and New York. However, they have relatively low per-capita carbon emissions. The reason is that some cities with high population density use an efficient and widespread public transportation system. But, in general, the highest carbon emissions due to transport are in the cities with high GDPs, and, conversely, the cities with the lowest GDPs tend to have lower emissions (ARUP/C40 Cities Climate Leadership Group, 2011).

If C40 mayors worked together toward a common goal, they could make a substantial impact on reducing environmental damage, health effects, and the deterioration of urban systems. They have the power and influence to mitigate the consequences of GHG emissions from transportation and other sources (ARUP/C40 Cities Climate Leadership Group, 2011).

## Cities' Contribution to Climate Change

Mayors who have control of city budgets and resources can act to have a relatively immediate impact on initiatives. Some cities have already taken initiatives and, depending on their success, other cities will follow. In cities where mayors have neither the power nor the resources to make changes and initiate projects this is not true. However, if mayors have control of municipal assets such as roads and pavements, they can move forward with transformative initiatives via direct projects and programs. Accordingly, these programs are often backed up by regulations and policies. Infrastructure initiatives led by mayors demonstrate they are capable of using their powers efficiently and promptly to collectively accelerate emissions reduction (ARUP/C40 Cities Climate Leadership Group, 2011).

The urgency for cities to mitigate a changing climate is more crucial today than ever. Some cities' contributions to GHG emissions are greater than that of many nations together. New York City, for example, emits 63.1 metric tons (MT) $CO_2$ equivalent to climate change. This number is only 10% lower than the contribution of all of Ireland's GHG emissions (Karabag, 2011).

However, not all cities release the same amount of emissions, because transportation, waste management, urban systems, land use, and technology differs among them (Karabag, 2011).

GHGs create considerable risks to the climate, and reducing those emissions can significantly impact the health of urban populations. Most experts advocate minimizing individual carbon footprints and investing billions of dollars to reduce the risks of radical changes in the earth's environment (Glaeser and Kahn, 2009). Combined evidence of anthropogenic effects on climate shows that the leading net annual flow of carbon released to the environment comes from $CO_2$ emissions from fossil fuel combustion (Braconnot et al., 2007).

The data set in Table 5.2 is derived from the following sources: IEA (International Energy Agency), CDIAC (Carbon Dioxide Information Analysis Center), EPA (Environmental Protection Agency), EIA (Energy Information Administration), and EC-JRC/PBL (European Commission, Joint Research Centre/Netherlands Environmental Assessment Agency). For comparative purposes the latest version of the data set includes greenhouse gas data through 2008 for three major countries with densely populated cities (World Resources Institute, 2012).

Analyzing carbon inventory for megacities is a crucial step in leveraging the capacity to reduce GHG emissions. However, megacities do not have identical emissions. Targeting the main sources of emission after determining their origins is key to effective implementation. The strategy can target four

**TABLE 5.2**

GHG Emissions for United States, China, and Brazil

| Region | Year | GHG Emissions Total | GHG Emissions Intensity | GHG Emissions per Capita |
|---|---|---|---|---|
| World | 1990 | 30017.6101 | 851.26 | 5.69 |
| USA | 1990 | 5978.9 | 750.22 | 23.95 |
| China | 1990 | 3593.9 | 2876.39 | 3.17 |
| Brazil | 1990 | 691.5 | 643.99 | 4.62 |
| World | 2000 | 33187.76504 | 694.63 | 5.46 |
| USA | 2000 | 6824.6 | 611.10 | 24.19 |
| China | 2000 | 4818.5 | 1430.65 | 3.82 |
| Brazil | 2000 | 938.2 | 680.05 | 5.38 |
| World | 2008 | No data | No data | No data |
| USA | 2008 | No data | No data | No data |
| China | 2008 | No data | No data | No data |
| Brazil | 2008 | No data | No data | 394.40 |

*Source:* World Resources Institute (WRI), 2012, Climate Analysis Indicators Tool (CAIT), version 9.0, Washington, DC: World Resources Institute.

*Note:* Emissions, million metric tons $CO_2$ equivalent ($mtCO_2e$); per capita emissions: metric tons of $CO_2$ equivalent ($mtCO_2e$) per person; emissions intensity: metric tons of $CO_2$ equivalent per million international dollars ($mtCO_2e$/million \$intl).

different categories: buildings, transportation, energy supply, and municipal services (Glenn, 2010).

Reducing a city's GHG emissions requires a set of strategic plans that, integrated toward a common goal, can be very effective in minimizing carbon emissions. Strategic steps to reduce GHG emissions such as energy efficiency in buildings and reducing traffic and car use on the roads are effective in lessening a city's carbon footprint (Glenn, 2010). Strategies for cities to develop and implement their infrastructure in a sustainable way will be crucial to combating climate change and will determine the future path of its success (Pennell et al., 2010). A city's commitment to reduce GHG emissions needs to be reinforced by an aggressive and informed strategy.

## The Hestia Project

The scientific understanding of carbon exchange with the land surface is crucial to creating an effective carbon monitoring system (Gurney et al., 2012). Utilizing top-down or bottom-up methods are efficient ways to create results of mitigation studies. Bottom-up studies are based on assessment of mitigation options and they target the usage of energy and greenhouse gas emitting equipment (for example, power-generating stations or vehicle engines). Finally, policy measures are included to create results (Al-Moneef et al., 2001).

The "Hestia Project" is the first research effort that uses bottom-up methods to quantify all fossil fuel $CO_2$ emissions, including individual buildings, road segments, industrial infrastructures, and industrial facilities for the generation of electric power on an hourly basis for an entire urban landscape (Gurney et al., 2012).

Combining a series of data sets and simulation tools, such as a building energy simulation model, traffic data, power production reporting, and local air pollution reporting, this method can be utilized in any major city in the United States and around the world. It can be applied as a principal factor in monitoring carbon emissions and implementing valuable greenhouse gas mitigation and planning. When measuring levels by zip code, the Hestia team attains a bias-adjusted Pearson $r$ correlation value of 0.92 ($p <$ 0.001) (Gurney et al., 2012). The Pearson $r$ measures the linear relationship linking two interval/ratio level variables. "The stronger the association of the two variables the closer the Pearson correlation coefficient, $r$, will be to either +1 or −1 depending on whether the relationship is positive or negative, respectively" (Pearson product moment correlation, 2012). It is crucial to interpret the global carbon cycle effectively through quantitative measures. This is a principal factor in advancing our understanding of climate change and its projections, and in expanding our understanding of ecosystem level biogeochemical principles to resourcefully manage climate change detrimental effects (Gurney et al., 2009).

---

## List of Terms

**Carbon dioxide ($CO_2$):** The primary greenhouse gas emitted through human activities.

**International Energy Agency (IEA):** An autonomous organization that acts to ensure reliable, affordable, and clean energy for its 29 member countries, including the United States, Canada, and many European nations. The IEA has four main areas of focus: energy security, economic development, environmental awareness, and engagement worldwide.

# Section II

# Risk Management and Its Relation to Climate Change

# 6

## Management Strategy for Effecting Change within Megacities

### Addressing the Full Spectrum of the Megacity's Risks

Risk management serves to identify potential opportunities, and then manage and take action to prevent adverse effects. It also emphasizes the probability of events and their consequences, which are measurable both qualitatively and quantitatively. These characterizations also apply to climate change risks and planning evaluations (Jones and Preston, 2010). A risk management framework addresses the full spectrum of challenges in areas such as planning, strategy, operations, finance, and governance. It also recognizes the specific needs of the megacity's different departments and functions as well as the potential impacts of parallel threats and events. The systematic analysis and management of health risks through a well-planned strategic approach to integrating recovery measures, preventing and mitigating risks, and identifying a population's vulnerabilities are a priority for the health sector to lessen public health risks impacts.

### Assessing and Initiating Steps to Manage Climate Change Risks with ISO 31000

The British Standardization Institution defines a standard as "a document established by consensus and approved by a recognized body that provides, for common and repeated use, rules, guidelines, or characteristics for activities or their results, aimed at the achievement of the optimum degree of order in a given context." Standards should be based on the consolidated results of science, technology, and experience, and aimed at the promotion of optimum community benefits (BS 0:2011, 2011). ISO (International Organization for Standardization) is a worldwide federation of national standards bodies with 119 members. The results of ISO's work are published in the form of international standards (Temmerman, 2000).

Currently, megacities are struggling to address the risk of climate-induced disasters. A methodology for developing and implementing the Australian/ New Zealand Risk Management Standard—AS/NZS ISO 31000 on the effects of climate change on populations would enhance the local and national capacity for effective actions. ISO 31000 supports a unique management approach to initiating steps that manage risk more effectively and developing a cohesive method to presenting adaptive strategies across cities around the world. International guidelines can help cities plan how to recover from sudden external risks (e.g., intense storms, earthquakes, tsunamis) and the more gradual external risks (e.g., sea-level rise, pandemics, and shifting disease patterns). Adaptation and mitigation models that confront the effects of a changing climate on health in urban populations, particularly on the poor, the most vulnerable to climate variability, are crucial to minimize risks of a disaster. This then becomes a fundamental component of frameworks for climate change adaptation and mitigation programs. While organizations have been conducting risk assessments for years, many still find it challenging to obtain their real value. A strong business case that applies Risk Management ISO 31000 and the business continuity management system standard (ISO 22301:2012) and recommends a systems view of risk assessment and proactive approach to risk management through a shared response at local and international levels would become increasingly important to measuring climate change. The resilience of megacities and sustainable development for communities and disaster risk management are addressed utilizing British Standard 65000 (Organizational Resilience) and the ISO 37101 Sustainable Development of Communities frameworks. They establish a common practice for using, creating, interpreting, and analyzing the city's operations while confronting its vulnerabilities and disaster-risk management strategies and presenting a solid foundation for continuous improvement. This enhances crisis management and business continuity management practices by integrating these into a wider resilience plan.

---

## Establishing the Context and Setting Objectives of Risk Management Assessments

To set the objectives of a risk management assessment, the location and scope of the study and the operating processes in the area under threat have to be established. An assessment seeks to identify and investigate risks to describe the actions that are required to attain the following objectives:

1. The potential impact on (i.e., damage to)
2. A particular value (e.g., house) from
3. A threatening process (e.g., waves, sea level rise) (Rollason et al., 2011)

As a consequence, the economic, social, and environmental values of the land where the strategic framework applies must also be included in the assessment. Risk criteria are also important considerations for the framework. Likelihoods and consequences scales and their combination in the current conditions must be included when defining the acceptable risk level. City leaders and stakeholders should determine what that level is and then identify tolerable and intolerable risks (giving priority to the latter) that need to be addressed according to local conditions (Rollason et al., 2011).

Risk modeling techniques are increasingly used by many governmental entities to evaluate exposure to natural disasters and crises. Assessing and comparing different types of risks from adverse impacts becomes fundamental in understanding and minimizing their impacts and recovering quickly after the event's occurrence on the balance of probabilities. There are three central risk modeling technique inputs:

- Hazards—Climatic events impacting a territory can present uncertainties about its hazard occurrences and the affecting area exposure. Existing knowledge of past events on a local or global context afford a tangible concept of the intensity and frequency of what is expected.

- Exposure—Location and geographical distribution of the territory that will be affected by a hazard need to be mapped. An account of human occupation and physical assets has to be performed. Particular components (e.g., geometric shape of exposed elements, economic value, human occupation and location) must be identified to differentiate the exposure.

- Vulnerability—Exposed elements susceptibility or probable conduct is directly linked to the level of hazard. Natural hazard can be differentiated and analyzed by the intensity and the frequency of the impacts. Hazard parameters can be established by analyzing and understanding past records of events and risk modeling probabilities. The nature, magnitude, frequency, and intensity can help determine the level of the hazard (World Bank, 2012a). Consequently, hazard models, which are based on a set of assumptions that should be conveyed to the model user, may present a reasonable account of complex environmental dynamics and evolution addressing many factors such as urbanization and climate change.

Socioeconomic trends assist populations and can also show their exposure to vulnerabilities and provide a clearer picture of the risks. Climatic events such as severe weather events and climate change affect hazard characteristics. Therefore, analysis of the long-term socioeconomic scenarios of the megacity is essential for an understanding of the probabilities under a risk assessment of vulnerabilities, while considering important factors such as urbanization and economic growth, lifestyle changes, and demographics (World Bank, 2012a).

## Principles of Risk Management: The Process of Creating Value

Creating a culture where the city's workforce addresses risk in every activity is crucial for effective plan development, implementation, and preparedness for any event. It builds common terms and metrics for addressing risks, increasing the chances of efficient actions toward climate change impacts on their population without adopting an unwarranted risk-averse position. The 11 principles of risk management, as defined in ISO 31000:2009, need to be considered in this process. This standard helps to accomplish the following goals:

1. Create and protect value.
2. Assure risk management is an integral part of all organizational processes.
3. Shape decision-making at all levels in the organization.
4. Manage uncertainty.
5. Ensure systematic, structured, and timely responses.
6. Take advantage of the best available information.
7. Make sure risk management is tailored to each organization.
8. Account for human and cultural factors.
9. Assure transparency and inclusiveness.
10. Provide a dynamic, iterative, and responsive framework to change.
11. Facilitate continual improvement of the organization.

Risk management frameworks, and specifically ISO 31000, can be highly useful in educating city leaders on how to minimize climate change and the effects of climate variability's public health risks effects on their city's populations. One specific major issue with climate change is trying to distinguish the zealotry from the reality of what we can actually achieve. The first step is to critically examine the objectives that are being set. Are they in fact achievable or are they a desirable pie-in-the-sky hope? Will they actually provide meaningful outcomes or are they essentially feel-good images? This is where ISO 31000 applies the process for testing the objectives and developing an effective understanding of what uncertainty exists and how or if it can be controlled within the resources and knowledge available to the program.

ISO 31000 provides the information needed to establish the principles, develops the right changes to the management system to be employed, and provides a process for understanding and managing the risks that will imperil the achievement of our objectives with respect to protecting the population from aspects of climate change. It also helps to pinpoint the factors we can change and those we cannot, as well as to know the difference between the two (Knight, 2013).

## Different Stages of Risk

Risk management can identify the different stages of risk and develop measurement processes for climate change. The framework to identify risk is recursive and can be periodically evaluated. The Intergovernmental Panel on Climate Change (IPCC) has identified and approved four generations of risk management:

- First and second generations investigate the nature of climate change concerns, and focus on identifying and analyzing risks and impacts.
- The third generation examines the nature of adaptation.
- The fourth generation implements the methods of evaluation and risk management.

Also, a fifth generation might be considered for researchers and stakeholders that would highlight the advantages of an adaptation and would develop frameworks to assess such benefits for improvement (Jones and Preston, 2010). The following steps are important to selecting risk management processes to evaluate climate vulnerability, climate change impacts, and adaptation strategies:

1. A scoping exercise where the context of the assessment is established. This identifies the overall method to be used and establishes relationships between stakeholders and researchers.
2. Risk identification. This step also identifies scenario-development needs.
3. Risk analysis, where the consequences and their likelihood are analyzed. This is a well-developed discipline with many methods available to undertake impact analysis.
4. Risk evaluation, where adaptation and mitigation methods are prioritized.
5. Risk management or treatment, where selected adaptation-mitigation measures are applied.
6. Monitoring and review, where measures are assessed and the decision made to reinforce, reevaluate, or repeat the risk assessment process (Jones and Preston, 2010).

## Analyzing Risk

The assessment framework is developed to methodically analyze the risks. Then, risk is evaluated by identifying the consequences and probabilities in the context of existing controls. Taking into consideration the source of risk is important and it has to be emphasized to understand the positive and negative effects and probabilities. An individual analysis of the consequences and likelihoods is needed to perform a qualitative climate change risk assessment based on particular outcomes. Analyzing past events, practice and relevant information, and applicable published literature is underlined for an effective analysis of risk (State of NSW and Office of Environment and Heritage, 2011).

The states of risk analysis stages are as follows (State of NSW and Office of Environment and Heritage, 2011):

1. Analyze existing controls—Identifying existing controls to minimize the consequences and likelihood of each risk. Only existing controls that are funded and completed should be measured in this stage. Assuming the climate change scenario is occurring, reducing the consequences and likelihoods of risk should be evaluated.

2. Analyze the event's magnitude of consequence and likelihood—Determine the phase of the magnitude of an event's consequence and its likelihood of occurring. Particular consideration should be paid to climate change scenarios and the existing controls to manage the risk.

3. Assign the risk priority rating—Using the risk priority shown in Table 6.1, the risk rating can be obtained. The process of analyzing different climate change scenarios has to be put into practice for each risk.

ISO 31000:2009 identifies the level of risk, expressed in terms of the likelihood of an event and its consequence. A description of the likelihoods and consequences to define the level of risk is fundamental to developing a risk management strategy for climate change (Rollason et al., 2011).

ISO 31000:2009 classifies risk evaluation as a framework to compare the results of the risk analysis with risk criteria and to determine if the level of risk is acceptable, allowable, or intolerable. The priority is given to intolerable risks. It is impossible to treat every risk, and there is a possibility that high implementation costs might offset the benefits or risk reduction achieved. The methods of reducing risks are evaluated and the actions to investigate new

**TABLE 6.1**

Risk Priority Ratings Example (Given That a Scenario Arises)

| Likelihood | Consequence | | | | |
|---|---|---|---|---|---|
| | Insignificant | Minor | Moderate | Major | Catastrophic |
| Almost certain | Medium | Medium | High | Extreme | Extreme |
| Likely | Low | Medium | High | High | Extreme |
| Possible | Low | Medium | Medium | High | High |
| Unlikely | Low | Low | Medium | Medium | Medium |
| Rare | Low | Low | Low | Low | Medium |

management measures are put in place. The integration of the likelihood and consequence in the previous step presents the "unmitigated risk": risks that are not diminished or moderated in intensity or severity. After implementing existing management measures in the assessment, identifying risk priorities that need immediate attention take place (Rollason et al., 2011).

## Treating the Risk

Risk treatments can be regarded as one type of climate change adaptation. These treatments can follow a series of principles that will positively and efficiently impact the implementation of the framework:

- Keep a balance between climate and nonclimate risks.
- Focus on high-priority risks to support the adaptation effort.
- Implement small, flexible, and incremental changes based on regular monitoring and revision of plans.
- Keep options for new adaptations open where possible.
- Focus on cost-effective actions.
- Review treatment strategies.
- Review existing risk controls to determine if existing controls are not sufficient.
- Identify changes in thinking or new measures to overcome gaps (State of NSW and Office of Environment and Heritage, 2011).

## Ongoing Monitoring and Review

Information about climate change is continuously updated. Therefore, the risk assessment process includes monitoring and review as an important part of the framework. This process should include the following steps:

1. Obtain new climate change information as it becomes available
2. Check that controls are effective
3. Assess new information obtained from events
4. Account for any changes in the process
5. Identify new risks and take action (State of NSW and Office of Environment and Heritage, 2011)

### Risk Assessment Framework Concerns

This volume proves that implementation of adaptation and mitigation risk-management frameworks for climate change has many benefits. Concerns and questions, however, remain regarding the challenges for many cities in executing risk management plans. Further research and policy development is necessary to address those issues in order to prevent potential risks due to climate change (Fifth Urban Research Symposium, 2009).

These actions raise ethical questions about the level of government involvement, what combination of stakeholders should prioritize the initiatives, and, consequently, the implementation of climate change strategies. Considering that awareness of the effects of a changing climate is low among vulnerable sectors of society and that the uncertainty of issues related to climate change at the local level is still high, cities face challenges in addressing the specific needs of the most vulnerable sections of its inhabitants: the urban poor (Fifth Urban Research Symposium, 2009).

Developing an assessment considering all factors of climate change impacts and potential risks, in addition to adaptive resilience and mitigation strategies that are not mere recommendations in reports but instead direct tangible actions toward the problem, is vital for the success of any risk assessments framework. Clearly, the absence of a city's climate change strategy will exacerbate climate change impacts, accentuating the risks to the urban poor (Fifth Urban Research Symposium, 2009).

Crafting a flexible and calibrated approach for building resilience and preparing social institutions for public health or disaster management to be retrofitted to adapt to climate change are important elements for city leaders to employ. Strategies that work for vertical coordination among national and local policy efforts must be emphasized. Promoting horizontal collaboration among a city's different departments and between global cities in a joint action has to be encouraged to get results (Fifth Urban Research Symposium, 2009).

### Financial Risks of Climate Change:
### The Importance of Risk Management

The choices many cities make today will reflect the outcome and degree of climate change impacts on populations and infrastructure in the future. There will be severe consequences if a city's government officials, leaders, and stakeholders do not take urgent action. The quantification of costs is important to assessing and developing a risk management plan, and insurance can help in that regard (Mistry et al., 2005).

Climate change could increase both the average annual losses and risk-capital requirements of insurers, which would increase the risk premium. For the United States, under a higher emission scenario, the risk premium could increase by nearly 80% by the year 2080. The potential costs of climate change impacts on cities can be avoided if a comprehensive risk management plan is put into action. Therefore, financial costs of climate change must be integrated in risk management to determine the proper actions for the future (Mistry et al., 2005).

## List of Terms

**Risk management:** The forecasting and evaluation of risks together with the identification of procedures to avoid or minimize their impact.

**Risk modeling:** The use of formal techniques to determine a risk.

**Risk treatment:** A step in the risk management process where risks are identified and risks that are not acceptable must be determined.

# 7

## Turning Theory into Reality: The Benefits of a Risk Management Approach

### Preparing the Risk Management Assessment against Climatic Impacts

The following list presents a set of facilities and systems identified to prepare the risk management assessment against climatic impacts (adapted from *NASA 2014 Climate Risk Management Plan*):

- Strategic analysis—How short- and long-term climate impacts will affect the megacities' operations. The risks associated with the impacts are categorized into the crucial groups listed later with examples of current and anticipated risks associated with climate change.

- Technical capabilities—The megacity's technical capabilities is vulnerable by climate vulnerability and change impacts. This risk is predicted to accelerate due to both sea level rise and the intensity and frequency of storms. Furthermore, the megacity's facilities are also at risk, several of which are especially vulnerable to coastal flooding due to sea level rise and storm surges as well as rising average temperatures and extreme events, such as heat waves and intense precipitation. For the megacity, leadership is becoming important to acquire and provide data and knowledge to inform decision making. This capability may be threatened by impacts to physical facilities. City employees supporting the development and delivery of data and knowledge are essential to minimize the impacts of a climate event. Impacts to supply chains may present risks to the city's capabilities and accesses, particularly if access to specific materials or chemicals is affected, which possess a high risk of environmental and health risks to the population.

- Built systems—The susceptibility of many systems across the city's infrastructure is exposed during a severe weather event. Energy, communications, and information technology systems are vulnerable to storms that are increasing in intensity and frequency. Heat waves present another detrimental effect where electrical blackouts and brownouts associated with these impacts threaten commercial energy utilities that provide the power any city uses to receive, process, and archive data. Heavy precipitation may overwhelm storm-water carriage systems, creating flooding that affects buildings and transportation systems and limits the ability of the city workforce to perform the duties to maintain a continuity business approach for the city's operation.

- Workforce and communities—Health risks and safety impacts to the population may pose operational risks. The population health can also be detrimentally affected by worsening air quality, heat waves, or problems caused by disease-borne vectors. Moreover, the personnel providing the necessary support to the city's operations is at risk and will not be ready to operate under these circumstances.
- Natural systems—Wildfires are an increased threat to natural and built systems. Wetland losses due to increased storm surge impacts may affect the safeguarding effect that shields some of the facilities, communications, road and rail network, and housing developments in the coastal areas. Increasing rainstorms and fluctuating groundwater tables may mobilize contaminants and consequently put the residents at a health-risk state.

Sharing information among the different departments in the megacity's operational infrastructure should include a qualitative assessment of the megacity's exposure to short- and long-term risks. Short- and long-term outcomes need to be considered in addition to the risks associated with these climate change impacts. Regardless, there are key questions for a city mayor to highly consider in order to prepare a wide-ranging climate risk management plan. The potential questions are

- What are the principal climate variables that drive and constrain the transportation as well as the water and energy sector, health, and food availability?
- What are the principal impacts of the identified key climate variables and processes on transport, water, energy, health, and food?
- What are the interactions between various impacts and how do they contribute to a megacity's aerosol and greenhouse gases emissions?
- Is it possible to detect thresholds for an impact of climate variables on the urban subsystems?

The key climate hazards and potential impacts to megacities' assets and capabilities are described in Table 7.1.

**TABLE 7.1**

Key Climate Hazards and Potential Impacts

| Key Climate Hazards | Potential Impacts |
|---|---|
| More frequent and extreme high temperatures and humidity | Increased risk of heat-related ailments particularly outdoors, higher cooling costs, decreased utility reliability, damage to buildings |
| More frequent and intense droughts; seasonal shifts in water cycle | Reduced water availability, higher water costs, saltwater intrusion, groundwater changes |
| More intense precipitation events | More frequent flooding of low-lying indoor and outdoor areas |
| Sea-level rise | Loss of usable land, inundation of coastal ecosystems |
| More frequent and intense coastal flood | Coastal erosion, safety implications for events in surrounding communities |

*Source:* National Aeronautics and Space Administration (NASA), 2014, *2014 Climate Risk Management Plan: Managing Climate Risks and Adapting to a Changing Climate.*

## The Plan–Do–Check–Act Cycle

Continuity management plans apply the Plan–Do–Check–Act (PDCA) cycle. This involves planning, establishing, implementing, operating, monitoring, reviewing, maintaining, and continually improving the effectiveness of operations (Draft International Standard ISO/DIS 22313, 2011). Assessing the risks that affect the city's operation is critical to organize and prepare the different departments within the jurisdictions against the potential impact of climate change (i.e., flooding, fires, power outages, infrastructure collapse) that ultimately interrupt the city's ability to provide basic and immediate services such as medical care, transportation facilities, and safe drinking-water resources (Business Continuity Planning Toolkit City of Philadelphia, 2009).

The following are considerations for assessing a business continuity management plan:

- Historical—What types of drastic weather events have occurred in the city and its jurisdictions?
- Geographic—What could happen in your city considering its geographical location and vulnerabilities to climate change impacts?
- Structural—What could affect the functional ability of the city's operations?
- Human factors—Human-caused emergencies can result from inadequate training, supervision, or negligence. (Business Continuity, 2009)

## Societal Security: ISO 22301—Continuity Management Approach

What would happen if your city was involved in a severe weather event? Would the city survive the crisis? How would you ensure that your city endures the disaster? Business continuity management (BCM) is about preparing an organization to deal with disruptive incidents that might otherwise prevent it from achieving its objectives. The international standard ISO 22301 specifies requirements to plan, establish, implement, operate, monitor, review, maintain, and continually improve a documented management system to protect against, reduce the likelihood of, prepare for, respond to, and recover from disruptive incidents when they arise (ISO 22301: Societal Security, 2012). This international standard can significantly reduce risk, particularly when a lack of awareness and a concrete strategy in a continuity management plan after a disaster or impact is present. It also provides flexibility of implementation, identifying what is most relevant to minimize financial loss, deaths, infrastructure, communications, and the overall operation of the megacity.

The effects of climate change on cities will be handled by the corresponding emergency services. But do we know what will happen next? Managing a city's overall continuity of operation becomes a priority in any type of severe weather event, as do implementing and operating controls and executing strategies for treating those risks. Any number of events can bring the city to a grinding halt, and business continuity management planning will ensure municipal leaders will respond judiciously to the circumstances. A continuity management approach contributes to a more resilient society. The following key components are considered when developing and implementing a continuity management system plan:

- Policy
- People with defined responsibilities
- Management processes relating to
  - Policy
  - Planning
  - Implementation and operation
  - Performance assessment
  - Management review
  - Improvement
- Documentation providing auditable evidence
- Any business continuity management processes relevant to the organization (Draft International Standard ISO/DIS 22313, 2011)

## Resuming "Normal Operations" after the Disaster

There are ten important points of action that determine how the city will preserve or restore critical functions. The city's quick resumption of normal operations after a severe weather event will affect the entire city's recovery and the welfare of its inhabitants.

1. Establish an emergency planning team—Workers from all levels and departments must be included in the team, focusing on those with expertise vital to the daily city's operation.

2. Identify who is in charge—It is important to identify who is in charge during a disaster risk event and ensure that all employees know who that person or position is. Establish a procedure for succession of management if that position or leader is not available at the time of the climate impact.

3. Examine the city's operation and activities—Identify internal and external operations that are important for the recovery and continuation of the city's different departments.

4. Identify an alternate location—Important consideration is necessary to identify a different location to run the city and/or different jurisdictions within the city's operations. Develop collaboration and viable assistance with "like" departments or the Public Building Commission/Department to share facilities if necessary.

5. Plan for citizens with special needs—Always include in your plan a set of specifications to treat and meet the needs of residents with special needs or disabilities.

6. Evacuation plan—Develop an evacuation plan for every work address for your city's operation. It is important to identify how notice will be given: an alarm, intercom, phone call, siren, and so on. All city employees and staff must be aware of where to assemble in the event of an evacuation or emergency occurrence.

7. Shelter-in-place plan—Develop a shelter-in-place plan for every address of your city's operations. It is used in dealing with the climate impact outside of your location, such as flooding, tornado/storm, or hail damages, where you take cover within the structure. All staff members should know where the shelter is located and have an access plan to reach the location.

8. Communication plan—Identify how the city's employees and staff will be advised of the emergency plan and what communication devices will be used in the climate change impact event. List the communication tools in order of preference, emphasizing the most

effective way for communication to the least effective according to the type of climate change impact.

9. Emergency contact list—The emergency contact list identifies how to contact staff in the event of a severe weather event or an indirect climate impact, such as flooding or power outage. This will include the technical means where the staff may receive different types of communications in order to respond and lead actions toward restoring and aiding the city's operation and the well-being of residents.

10. Write a plan—Document and update your city's continuity management system plan at least once a year. (FEMA: Are You Ready? An In-Depth Guide To Citizen Preparedness, 2004)

It is difficult to predict or analyze the type of activities that will be interrupted due to climate change impacts in the megacity and how these impacts will affect the health of the population. A continuity management plan draws attention to the impact of disruption and identifies those activities where the megacity's operation needs to focus for its survival. A continuity management system plan assists the megacity's leaders in recognizing what needs to be done to protect its residents and infrastructure. The ISO 22301 business continuity management system may also be able to take advantage of opportunities that might otherwise be overlooked or considered to be too high risk (Draft International Standard ISO/DIS 22313, 2011).

## Adapting and Using the Framework

Effective response and recovery to climate change impacts involves taking prior actions before the event strikes. A proactive approach must be established along with planning for the likelihood of a weather event and its effects on public health that have the capacity to interrupt the city's operations and impinge on the well-being of the population. It is necessary to emphasize the personal commitment of the city leaders to dismiss the thought that "it won't happen in my city." Catastrophes happen on a daily basis and, as previously mentioned, preparedness means being proactive and planning. That is the essence of efficient business continuity planning (State of Queensland, 2009).

## Climate Change Impact Analysis

Megacities have overall functions that are important for maintaining a normal level of operations, but only a percentage of those operations are going to be vital for its continued normal level of functionality. City leaders need to gather information to determine basic recovery requirements in the event of a disaster. Identifying which areas and departments of operations of the city will be most affected by climate change and climate variability is crucial in order to develop a plan and to determine what effect it will have on the city as a whole (State of Queensland, 2009). The Climate Change Impact Analysis (CCIA) will be used to establish the city's critical business activities, the resources required to support such activities, and the impact of ceasing to perform these activities.

## Recovery Time Objectives, Planning, and Strategies

Recovery time objectives (RTOs) need to be part of your climate change impact analysis and assigned to each activity. It identifies the time from which you declare a crisis or disaster to the time that the critical business activity must be fully operational in order to prevent serious infrastructure damages, health risks, financial losses, and loss of lives (State of Queensland, 2009).

The response phase of the business continuity planning takes place as the climate change impact event happens and immediately after it occurs. The plan takes actions to respond to the event suppression, minimizing the loss of lives and property, and ensuring that the climate event won't escalate into a chain reaction of incidents that will be more difficult to deter. Response to an incident generally involves a response from management, operational, and communications departments. An incident response management team mainly takes care of the response phase of the plan. Responding quickly to a negative impact on the city infrastructure, people, and so forth becomes vital to reduce negative impacts; although climate change events are often unpredictable, they are not always unexpected. If the impacts and effects of climate change are not dealt with in a prompt and effective manner, they have the potential to have detrimental repercussions on the city's financial situation, government, and population health (State of Queensland, 2009).

The actions taken to recover from a devastating climate event may not always be possible. These actions can be generally identified as

- Resumption (continuity) of business activities
- Restoration (recovery) of resources (State of Queensland, 2009)

Recovery planning is a proactive approach to restore operations in a prompt manner after a climate event hits the city. This is a crucial task to implement, because a serious incident can occur at any time. To get their personnel prepared and in the right mindset, mayors and the city's key decision-makers must ensure the staff is well trained and rehearsed throughout the planning process. These actions will ensure a better chance of recovery.

The recovery plan focuses on how to bring back important activities after the climate event has taken place. City leaders must assemble teams dedicated to the following tasks:

- Recovery team objectives—Must be defined and those goals need to demonstrate a clear understanding of the recovery planning that reflects what is needed to continue operating the city. A recovery team must be set up with the allocation of backups to ensure awareness of their functions in the recovery process.

- Disaster recovery—Establish a disaster recovery location where the city's vital staff may work off-site to be able to access significant backup systems, records, and supplies. Identify essential assets that need to be recovered and require special protection. Safety-storage areas off-site and fireproof cabinets are needed.

- Communication strategies—Develop strategies for communication with other organizations in the region affected (e.g., community centers, churches, schools) as well as other cities and states. This effort allows the city leadership to assess the impact area, exposure, and consequences, provide health care assistance, and mobilize the people affected by the event to allocate them to a safe area for treatment and temporary housing assistance.

- Planning for disruptions—Disruptions to electricity, water sewerage, and telecommunication systems are unavoidable after a severe weather event. Backup systems and other alternatives need to be available that can be used after the impact (State of Queensland, 2009).

## Creating a Focus on Execution and Progress

Some megacities, as well as cities with lower population densities, will experience physical changes. Infrastructure such as rail and road networks, communications, buildings, health centers and hospitals, in addition to the local and long distance transportation sector, delivery of medical aid, and other vital services will be interrupted as a result of extreme events. Higher temperatures may restrict the amount of time city workers and medical staff can safely engage in assisting urban residents with their urgent needs. On the other hand, climate change provides other opportunities for some cities such as reducing work stoppages caused by frost, prolonging the portion of the year during which construction, preparedness for emergency events, training personnel with risk management tasks and other beneficial activities are able to take place (Sussman and Freed, 2008).

In that context, adaptation strategies implementing the ISO 31000 risk management framework and ISO 22301 business continuity management planning may also create new product markets (e.g., climate proofing materials, building designs with an environmental concept for urban locations). Moreover, reallocation of market needs by making locally sourced materials in demand successively reduces travel costs, which minimizes carbon emissions (Sussman and Freed, 2008).

Financial risks associated with climate change physical effects are of a great concern when addressing mitigation efforts. The recent hurricanes, flooding, cyclones, heat waves, and other weather extremes around the world made many city leaders aware of the risks associated with them and created awareness of the potential direct impacts of severe climate events. However, this understanding of the detrimental impacts of climate change has not translated into tangible actions toward the long-term effects of changes in seasonal variation in temperature and precipitation that affects the city's infrastructure and the health of the population (Sussman and Freed, 2008). Most sectors of the megacity's economy have different vulnerability levels to the physical effects of climate change. Table 7.2 describes the sectors of the economy that are at risk and potential local repercussions. As the table shows, all sectors are affected (to different degrees) with the possibility of property damage, business interruption, and changes or delays in services provided by public health and private entities, electricity and water utilities, and transport infrastructure (Sussman and Freed, 2008).

Some cities are already taking into account climate change strategies because they are located in vulnerable geographic areas where climate changes are occurring in the intensity or frequency of extreme events. These cities are more likely to be more concerned and have a proactive approach toward the long-term effects of climate change and are to a certain extent responding with planned actions (Sussman and Freed, 2008). These actions are not lessening the life and financial losses as well as the overall health impacts and mental health issues witnessed after Katrina, Sandy, Haiyan, and other severe climate events of that magnitude.

Climate-related risks to property, health, crops, business interruption, and other activities have been regarded by the insurance industry as one of its priorities for many years. Many business case studies have shown benefits from a planned and proactive approach to extreme events not only by adapting to a short-term plan but also incorporation of systematic long-term planning into their business operations (Sussman and Freed, 2008). Table 7.2 describes some functions that might feel the effects of climate change.

**TABLE 7.2**

Areas at Risk Resulting from Physical Effects of Climate Change

| Areas of Impact | Physical Effects |
|---|---|
| Utilities and city operations | • Peak electricity demand due to warmer and more frequent hot days could in some regions exceed the maximum capacity of current transmission systems and will be combined with system stresses due to heat<br>• Increased risk of damage to facilities and infrastructure from extreme and unpredictable weather conditions<br>• Uncertainty over energy output from hydroelectric plants due to potential water shortages<br>• Uncertainty over water supplies for cooling power plants<br>• Extreme weather events increase physical risk to business operations, for example, due to flooding<br>• Risk of food supply and operations interruptions due to extreme weather events<br>• Longer-term weather trends may affect reliability (and quality) of supply of fresh produce<br>• Physical risk to water supply and raw materials<br>• Greater risk of animal infections (e.g., avian flu), insect infestation |
| Building design and construction | • Extreme weather events may disrupt transport for site deliveries and affect site work (e.g., muddy site conditions), restricting workdays<br>• Infrastructure (e.g., drainage) affected by extreme weather events<br>• Excessive heat in summer will affect some construction processes and onsite workforce<br>• Design standards may need to be clarified or upgraded in response to changing climate<br>• Insurance may be more expensive or difficult to obtain for existing buildings, new buildings, and during the construction process<br>• Increased need to develop catastrophe models to evaluate capital adequacy and overall natural catastrophe exposure<br>• Disruptions to business operations become unpredictable and more financially relevant<br>• Competition for water resources between agricultural and urban development increases commercial risks with impacts on crop insurers<br>• Increased risks to human health (thermal stress, vector-borne diseases, natural disasters)<br>• Prolonged periods of poor weather or extreme events increase costs of claims and make it more difficult to deal with high volumes of claims |

*(Continued)*

**TABLE 7.2 (CONTINUED)**

Areas at Risk Resulting from Physical Effects of Climate Change

| Areas of Impact | Physical Effects |
| --- | --- |
| Insurance | • Increased need to develop catastrophe models to evaluate capital adequacy and overall natural catastrophe exposure<br>• Disruptions to business operations become unpredictable and more financially relevant<br>• Competition for water resources between agricultural and urban development increases commercial risks with impacts on crop insurers<br>• Increased risks to human health (thermal stress, vector-borne diseases, natural disasters)<br>• Prolonged periods of poor weather or extreme events increase costs of claims and make it more difficult to deal with high volumes of claims |
| Food | • More refrigerated distribution and storage required and problems with livestock transportation in summer heat<br>• Damage to transportation infrastructure or disruptions in services due to, for example, floods, creating problems with transporting raw materials<br>• Limited availability of water and potential interruption of supply to irrigation systems<br>• Equipment and other investments, as well as expertise of farmers and workforce, are linked to specific crops, which may become unprofitable or may no longer be viable<br>• Quality issues: overheating of grain or availability of water for prewashed products<br>• Access to land during flood or extreme rain conditions<br>• Less frequent frosts will affect quality of certain crops and reduce kill-off of pests/disease<br>• Exposure of workforce to increased heat<br>• Farm buildings affected by extremes of wind, heat, rain (animal welfare issue) |
| Transportation/ manufacturing | • Supply chain interruptions and vulnerable transport systems carrying local products and services and/or high-value products around the world (e.g., local transport carrying medical supply, ship carrying over $60 million of products affecting and delaying the city's ability to recover after an extreme event due to losses in revenue)<br>• May need vehicles that tolerate new extremes of climate, including greater intensity of rainfall (affecting seals, wipers, tires) and increased need for cooling<br>• Process environment will become hotter with increased need for cooling, particularly important for comfort/health of workforce and performance of production processes |

*Source:* Sussman, F., and Freed, J., 2008, *Adapting to climate change: A business approach,* Arlington, VA: Pew Center on Global Climate Change.

## The Continuity of Health Care Systems

As business continuity management strategies place a core function in the preparation and the disaster recovery to resume operations, a network of coordinated health care services must be ready to take action to operate during and following the disaster. Health care systems are as effective and resilient as the communities and regions where they are located. The health system network needs to be available to provide services for emergency care, which is the safety net of the entire health care system, caring for everyone, regardless of the gravity and/or intensity of the injury. Therefore, individuals and communities within the megacity need to have the aforementioned services available even during extended utility outages and transportation infrastructure disturbances.

## Resilience and Risk Management

The foundation for increasing resilience is risk management. Increasing resilience involves a series of steps to understand, identify, and reduce risk (American Meteorological Society Policy Program, 2014). Hurricanes Katrina and Sandy in the United States have shown the fragility of the city's system operations and recovery strategy. As a result, health care delivery needs to be decentralized from hospitals to a range of subacute settings. These venues cover a range of medical care services that support the individual's continued recovery from illness or management of a chronic illness or disability such as assisted living, ambulatory facilities, and home care to be able to treat and provide short- and long-term care (U.S. Department of Health and Human Services, 2014).

The levels of resilience can vary according to the capability of the megacity to cope with those risks and for that instance community engagement is a vital element for health care systems to efficiently function. Disaster survival and recovery are intertwined with access to health care services to prepare for and recover from climatic events. Social factors influence the communities' competences to resilience, and health care organizations must compound efforts to understand and coordinate actions with particular hospitals that carry the capacity to function under overwhelming circumstances as well as residential care facilities, ambulatory, and home care programs, and the social and environmental justice issues that define their communities (U.S. Department of Health and Human Services, 2014).

## Setting the Stage for Action: Climate Change Confounds "Uncertainty"

The defining characteristics and complexities of climate change require a strong decision-making process, as megacities' leaders face unavoidable choices. International standards define *risk* as the "effects of uncertainty on the ability to meet stated objectives." All major cities have objectives that define their service to their citizens. But there are so many uncertainties that can hinder the ability to be able to meet these objectives. Climate change helps confound the degree of uncertainty for the megacities. There is, however, a problem in communicating the impacts to the cities. The "message" is not always clear to the governance of these jurisdictions. Climate change scientists use an arcane language with terms like vulnerability assessments and adaptation. Although they are very important steps for a resilient megacity, these concepts are not always actionable. They point out the problems and the dire situation, yet offer little in the way that a megacity can respond to the circumstances. City leaders need to move past the doom and gloom to some positive actions that can help deal with the situation (Pojasek, 2013b).

Today, economic and environmental systems are at risk of collapsing, and these challenges dissuade any initiatives to develop and implement concise climate change resilience frameworks. The economic difficulties being experienced globally are exhausting financial resources while putting city leaders in a difficult position, as the impact of climate change is more evident with rising temperatures and more frequent extreme weather events. Structural changes and strategic investments in megacities' operational infrastructure are crucial to call attention to the perils of climate change for the urban poor's health and well-being. As reported by the World Economic Forum's Quarterly Confidence Index, as well as the Global Risks Perception Survey, major systemic financial failure is the economic risk of greatest systemic importance for the next 10 years. The failure of climate change adaptation and mitigation strategies are also considered to be urgent within the next decade. In comparison to previous surveys, the lack of a climate change adaptation strategy replaced rising greenhouse gas emissions as the most systemically crucial plan for action (World Economic Forum, 2013).

Public leadership is a demand for megacities, as many stakeholders are open to the benefits of public–private partnership. Therefore, stronger governance needs to be implemented as stakeholders predict an emphasis on public ownership. The Organisation for Economic Co-operation and Development (OECD) report that studied infrastructure challenges and governance trends shows that stakeholders predict an emphasis on public ownership. It is highly important to create strategic solutions at a city-wide level as cities move to greater central control and autonomy within municipal government. The private sector can manage and increase the efficiency of services, providing strong leadership for the government and public sector (Megacity Challenges, 2006).

---

## Risk Strategies: Private and Public Sector Linkage

*Kurt J. Engemann and Holmes E. Miller*

### Risk and Resilience

Society depends on essential resources being available in order to function properly; however, disastrous events can disrupt critical processes, causing severe problems. Although planning for disasters is often overlooked by those who are unaware and otherwise preoccupied, prudent administrators develop risk strategies to reduce the probability of such events and to lessen their impact. Risk strategies are used to manage possible events that potentially have very large negative outcomes. The appropriate strategy to implement depends on the likelihood of the event occurring and the magnitude of the potential loss. Business continuity management, in the private sector, and emergency management, in the public sector, are increasingly recognized as core activities to provide resilience for organizations of all types and sizes. This section discusses issues relating to risk strategies and associated linkages between private and public organizations.

### Private Sector

Identifying potential events that threaten an organization and building resilience for an effective response requires actions before, during, and after a crisis. Business continuity management is a holistic management program that identifies potential events that threaten an organization and provides a framework for building resilience. This comprehensive program should span the entire organization, and its top priorities should be safeguarding human life and protecting the environment. The most cost-effective strategies that accomplish business continuity objectives must be kept current and properly tested to ensure that the appropriate measures are taken in situations requiring activation. Commitment must be secured from senior management who determine organizational objectives.

A business impact analysis identifies the importance of the organization's activities by assessing the impact of their interruption. To build resilience to threats, it is necessary to understand the organization's deliverables and what resources are essential to ensure continuity of critical activities. The organization's activities are assessed and the impact over time of their inter-ruption provides the rationale upon which appropriate continuity and recovery strategies are formulated.

Organizations require resources including personnel, facilities, equipment, inventory, utilities, and systems, and disruptions can occur if important resources are unavailable. A recovery time objective is the prospective point in time when an operation must be resumed before a disruption compromises the ability of the organization to achieve its objectives. An event may be designated as a disaster if the disruption continues beyond the recovery time objective. A business impact analysis determines the resource requirements to enable each function within the organization to achieve continuity and recovery within established time frames.

All critical areas of the organization that must remain operational or recover promptly need to be identified. Critical operations include life-safety and environmental controls, as well as customer service and revenue-generating operations. Information technology is a support function that is usually time critical. Some organizations require general office space that is easily replaced, whereas some organizations need buildings with special systems that cannot be readily restored. The destruction or denial of access to buildings could disable normal operations for a period of time. Building contents such as furniture and office supplies are easily replaced, however, specialized equipment can be more problematic. Organizations that have valuable, unique equipment should have protection measures in place.

In addition to internal resources, organizations have external dependencies including suppliers of materials and services. Special attention must be placed on any single-source provider of critical items. The 2011 earthquake and tsunami that struck Japan, a vital supplier of parts for major industries, had a ripple effect worldwide causing temporary shutdowns because of the lack of Japanese parts. Essential community infrastructure services are also vital, including fire, police, transportation, and utilities (communications, electricity, natural gas, sewer, and water).

It is necessary to understand the impact of disruptions over time by identifying critical functions and operations, identifying potential impacts and determining when impacts begin. A business impact analysis provides direction for business continuity management by providing a foundation for the program. Information that is assembled helps identify assets and processes that require protection, defines recovery time frames, determines resource requirements, and provides a basis for strategy development.

The inability of the organization to provide services has both a direct and indirect financial impact. Insurance can provide some coverage of financial losses associated with downtime and the destruction of organizational assets, however, insurance does not cover most indirect losses. Losses resulting from a crisis increase the longer operations are below usual levels.

A risk event chain describes the transition from threat to crisis to disruption to impact. For example, a flood is the threat and a crisis would be the event of a flood affecting a particular facility. The flood can destroy assets and cause a disruption at a facility resulting in a loss of revenue. Controls can reduce the probability of transitioning through the risk event chain and can mitigate the resultant impact. It is possible for different crises to result in the same disruption; for example, a data center could be destroyed by a flood, fire, or explosion. Because identifying all possible crisis events is impractical, the events selected for analysis should represent the most significant exposures confronted by the organization.

Risk is the possibility of experiencing an event, measured in terms of probability and impact. The purpose of a risk assessment is to prioritize planning by evaluating the likelihood of events and their potential impact on critical functions. Risk assessment begins with a methodical process of identifying threats and determines the risk that each threat poses to the organization. A threat is a source of potential negative impact that may manifest in a crisis. If not managed well, a crisis may have a severe undesirable impact. A disaster is a major crisis that imperils an organization. An event may be deemed to be a disaster due to factors such as loss of life, environmental damage, destruction of assets, and duration of disruption.

Threats are always present and signify possible sources of detrimental impact to an organization. Threats can be natural, accidental, or man-made, and can lead to disruptions in operations that can adversely impact the organization. A risk assessment determines the most significant threats to the organization and should be focused on the most crucial business functions identified during the business impact analysis. The steps of a risk assessment are identify threats, evaluate controls, estimate probabilities, determine impacts, evaluate risks, and prioritize risks.

The severity of a threat depends upon factors including geography, infrastructure, and political and economic situations. A methodical approach to study threat data is to commence with a wide view and then continue to a detailed view. For example, threats can be identified in the general region (e.g., hurricane), in the community (e.g., power outage), and in the building (e.g., fire).

Controls are procedures and devices that inhibit the occurrence of a crisis or mitigate its impact. Controls include preventive maintenance, physical security, information security, and personal procedures. The evaluation of controls includes determining the benefits of the controls, identifying costs, and improving the controls. Controls can improve workplace safety, reduce the probability of an event, and reduce losses by providing some business continuity. The annualized cost of a control is compared to the reduction in the expected annualized risk exposure. Some decisions are clear, whereas others require more detailed quantification. Improvements to existing procedures and physical controls to mitigate damages, injuries, and loss of life should be identified. Organizations with hazardous materials require special attention to avoid environmental contamination and health hazards.

Employees are critical to the success of any organization, and safety and security issues are vital. Subsequent to a community-wide crisis event, the recovery of the workforce is a high priority. After a community-wide crisis event, such as a hurricane, employees may not be able to work for a number of reasons. Even if an organization's facilities remain intact, residential damage, breakdown of community infrastructure, and unavailability of gasoline can keep the workforce isolated and unable to commute to work.

Hurricane Sandy, the devastating East Coast storm and its aftermath, killed more than 100 people, devastated communities in coastal New York and New Jersey, left tens of thousands homeless, crippled mass transit, triggered paralyzing gas shortages, inflicted billions of dollars in infrastructure damage, and cut power to more than 8 million homes, some of which remained dark for weeks.

Estimating the probability of crisis events involves reviewing historical data and discussing the events with relevant groups such as the fire department, weather bureau, utility companies, police departments, engineers, and government agencies. It is important to consider risk factors including weather, topography, population, transportation, infrastructure, facilities, and supply chain. Judgment is often used when data are unavailable.

The level of impact of a crisis event may be based upon various criteria such as loss of life, environmental damage, destruction of assets, and duration of disruption. The impact of an event includes downtime of a critical resource. For example, a severe winter storm might cause little damage but completely shut down all organizations in a given area for one day. For organizations with time-sensitive operations (e.g., a hospital) the loss of a day is important. Over 200 patients at New York University Medical Center were evacuated after power went out as a result of Hurricane Sandy and generators began to fail. For some other organizations (e.g., a school), a one-day loss of operations will have little impact.

Disruptions in operations can result in losses, both direct and indirect. Direct losses include losses due to physical damage, expenses related to incremental personnel, and losses resulting from failure to process deliverables in time. Indirect losses are permanent losses of future business due to the disruptions, for example, customers who switch their business to competitors. The expected annualized losses for an event are the sum of annualized losses for all areas affected by the event, maintaining that common losses are not double counted.

There are several methods of classifying risks, but most approaches use some risk measurement involving event probabilities and impacts. This score can be used to prioritize risks. An organization develops a strategy to manage risk through avoidance (eliminate risky activity), transfer (reassign risk), reduction (reduce likelihood/impact), and acceptance (retain risk). Strategies may prevent disasters, provide backup for lost resources, or specify alternative resources. A strategy may provide protection against only one event or against many events. Continuity strategies increase the organization's capability to respond to events in order to continue operations, and recovery strategies improve the organization's capability to return to stable operations following an event. Risk strategies include

- Alternate sites
- Backup equipment
- Buffer stock
- Building fortification
- Distributive processing
- Distributive warehousing
- Insurance
- Manual overrides
- Maximizing work hours
- Multiple suppliers
- Mutual aid agreements
- Outsourcing
- Personal protection equipment
- Portable equipment
- Prepositioning
- Quick resupply
- Reciprocal agreements
- Redundant operations
- Remote-site storage
- Social distancing policies
- Subcontractors
- Temporary structures
- Work from client location
- Work from homes and hotels

Some crises have associated warning periods, in which case it may be possible to preposition employees and equipment to a safer location. A crisis can affect a large area, and in addition to direct organization damages, the normal community infrastructure support system may be degraded. Relocation of the organization within the community may be difficult or impossible in the short term. Occasionally, the added expenditure required for contingency planning is insignificant because it can be combined with other plans. For example, the decision to decentralize operations is made simultaneously as additional space is being acquired for growth.

Available strategies need to be evaluated for each critical function. Some strategies cover multiple events and cover several operations. Offsite strategies provide coverage for the broadest range of events, whereas onsite backup could be a cost effective and convenient solution for some events. The selection of a set of preferred strategies depends on costs and benefits.

A strategy involves both one-time costs and recurring costs. These costs are annualized to allow for comparison of strategies. One-time costs include the costs of facilities, equipment, computer hardware, furniture, staff, and consulting. Recurring costs include the costs of renting/leasing, maintenance, services, personnel, supplies, and utilities. The cost of a strategy increases as allowable downtime decreases. For example, a short time frame may require a standby building and backup equipment. The total annual cost is the sum of the annualized one-time costs and the recurring annual costs.

Quantifying the benefits of a strategy requires reviewing all the events for which the alternative provides some protection and estimating how it reduces the expected loss of each of these events. Estimation of losses should be done in the context of the expected response of competitors, the strength of customer relationships and the impact on image, and should be considered in conjunction with insurance coverage.

The benefit of a strategy is the reduction of potential losses compared to the base case. The impact of a crisis depends upon its extent and duration. The net benefit of a strategy is obtained by subtracting the total annual cost of the alternative from its benefit. Sensitivity analysis should be used to determine how changes in the estimates of costs, losses, and probabilities would affect the selection of a strategy.

Management selects strategies by reviewing and rating various combinations of strategies. The advantages and disadvantages of the strategies should be determined and the suitability of the strategies should be assessed against the business impact analysis. Some of the evaluation criteria include cost, service level, time to switch, reliability and manageability. Management judgment is used in both assigning the ratings and selecting the strategies.

Before selecting strategies, it is necessary to review the resources required to conduct operations and to assess the level of resources that are available. Selecting strategies involves reviewing business continuity objectives, assessing strategies, consolidating strategies across the organization, determining advantages and disadvantages, and conducting cost-benefit analyses. After senior management has authorized the strategies, arrangements must be made to implement them, which involve acquiring equipment, contractual agreements, and workplace changes, preparing backup and off-site facilities, and ensuring that appropriate documentation is in place.

## Public Sector

No business continuity plan exists in a vacuum. Private organizations' business continuity plans depend on public entities, and public entities' emergency management efforts depend on private organizations. The business continuity phases are prevention, mitigation, response, recovery, and restoration. Prevention actions lessen the likelihood of a crisis. Mitigation reduces the impact of an event. Response is the immediate reaction of an organization to an event. Recovery is the stabilization and resumption of critical operations. Restoration is the process of returning to normal operations at a permanent location.

Emergency management has four phases: mitigation, preparedness, response, and recovery. These phases are similar to phases in business continuity. But in the case of emergency management, mitigation refers to the impact of events that already have occurred rather than steps to minimize the damage from events before they happen, for example, building codes that may reduce loss from earthquakes or sand dunes to prevent flooding from hurricanes. Regardless of the definitions, one thread that is woven throughout all phases is the importance of relationships among organizations, both within the public sector, within the private sector, and between the two sectors.

For the public sector, three obvious levels of interdependencies involve those occurring among local, state, and federal entities. Added to this are nongovernmental organizations (NGOs) that frequently play a role when disasters occur, such as the Red Cross. Examples of parties involved in the interrelationships to be considered when planning for crisis events include police, fire, emergency medical services (EMS), and other volunteer agencies, at both the state and local level, in addition to federal agencies, such as Federal Emergency Management Agency (FEMA). Coordination of plans and resources among these various agencies is critical to the success of the response and recovery effort. This includes coordination in developing strategies for the plans, implementing and testing the plans, and ongoing maintenance of the plans.

Perhaps the most visible example of a disaster where coordination did not occur or when it did, occurred poorly, was the response to Hurricane Katrina. The executive summary of *The Report of the Senate Committee on Homeland Security and Governmental Affairs* provides a postmortem of what went wrong and what went right. The report states:

> But the suffering that continued in the days and weeks after the storm passed did not happen in a vacuum; instead, it continued longer than it should have because of—and was in some cases exacerbated by—the failure of government at all levels to plan, prepare for and respond aggressively to the storm. These failures were not just conspicuous; they were pervasive. Among the many factors that contributed to these failures, the committee found that there were four overarching ones:
>
> 1. Long-term warnings went unheeded and government officials neglected their duties to prepare for a forewarned catastrophe.
> 2. Government officials took insufficient actions or made poor decisions in the days immediately before and after landfall.

3. Systems that officials relied on to support their response efforts failed.

4. Government officials at all levels failed to provide effective leadership. These individual failures also occurred against a backdrop of failure, over time, to develop the capacity for a coordinated, national response to a truly catastrophic event, whether caused by nature or man-made.

While this conclusion alludes to the incompetence and poor performance of many officials, there also is a structural element as well. This is illustrated in the coordination problems among three key officials: New Orleans Mayor Ray Nagin, Louisiana Governor Kathleen Blanco, and FEMA Director Michael Brown. A PBS documentary on a postmortem of the event illustrates the finger-pointing by each toward the others, even after the fact, and in doing so illustrates the coordination problems that occurred during the hurricane itself.

For example, Michael Brown said, referring to Governor Blanco, that no one was in charge; the governor denied this but says that resources that FEMA should have provided—such as buses—were not available. Mayor Nagin said that National Guard resources that should have been ordered by the governor came late and were too few. Elsewhere, there has been much criticism of both the mayor's role and the performance of FEMA Director Brown. Beyond the finger-pointing lays a reality of poor preparedness, poor interaction, and poor execution of what plans existed. The silo effect of the various governmental units does not even address mitigation issues regarding deficiencies in the design and upkeep of the levees and the role of the United States Army Corps of Engineers. In many ways Hurricane Katrina serves as a textbook lesson in highlighting areas where poor leadership at every level, communication among various levels of government, and poor emergency management processes can be improved.

Hurricane Katrina also illustrates what can go right and in doing so, highlight lessons learned that others could follow. The Senate report states:

> The effective and heroic search and rescue efforts by the U.S. Coast Guard; and the outstanding performance of certain members of the private sector in restoring essential services to the devastated communities and providing relief to the victims. These successes shared some important traits. The Coast Guard and certain private sector businesses both conducted extensive planning and training for disasters, and they put that preparation into use when disaster struck. Both moved material assets and personnel out of harm's way as the storm approached, but kept them close enough to the front lines for quick response after it passed. Perhaps most important, both had empowered front-line leaders who were able to make decisions when they needed to be made.

These successes illustrate how an individual organization—such as the Coast Guard or a private company like Wal-Mart who also responded with supplies—can effectively respond to disasters. Why is this so? Leadership, foresight, effective planning, and execution are key reasons. Even for a significant crisis event, such as the case with Hurricane Katrina, the attributes exhibited by the Coast Guard and by private entities can be exhibited by others as well. In this case the Coast Guard played the role of providing resources to assist in the response and recovery efforts, which is part of its mission. Private companies such as Wal-Mart executed their own business continuity plans so effectively that they could also provide additional resources to support the plans of the city and state. In hindsight, a more effective emergency management plan would have incorporated private sector resources in the plan, rather than fortuitously relying on them after the crisis event occurred.

Just as intrasector dependencies exist in the public sector, private sector organizations depend on each other. Many organizations' business continuity plans recognize this. Examples include dependencies on telecommunications companies; backup sites for computer processing; prepositioning supplies; and backup agreements with other organizations that can provide computer processing power, space, and even personnel when disaster strikes. For example, a bank needs to align its business continuity plan with those of major customers, internal network, external telecommunications providers, and network service providers, such as the Federal Reserve. An automobile manufacturer's business continuity plan depends upon those of suppliers of electrical power; sea, rail, air, and/or truck transport; component parts suppliers; and suppliers of computing power and storage in an era of cloud computing. A hospital's business continuity plan depends upon electrical power, medical supplies, personnel from outsourced services, and ambulance services. In all of these cases, success of one organization's business continuity plan depends on the availability of services from others, which often means the success of their business continuity plans as well. Whereas people certainly are involved in private company business continuity plans, public sector emergency management plans by definition must deal with events involving many more people, which creates an additional dimension of complexity and a greater need to identify interdependencies and draw on resources to deal with them.

## Public–Private Sector Interdependencies

The interdependencies between the public and the private sector are pervasive. Often, they are self-evident and readily identified. For example, a city's emergency management plan depends on and should incorporate the plans of local telecommunications and health care providers. The business continuity plan for a financial services provider in New York City would—for a flood event—depend on the plans for the city's subway and bus systems.

Often, however, the dependencies are not as readily apparent, even when hindsight said they should have been considered. During Hurricane Katrina, buses were available but bus drivers were not. When asked why the National

Guard could not have supplied bus drivers, Mayor Nagin claimed, "They have to be activated by order of the governor. During our event, they were slow coming also. We had maybe 200 to 250 National Guardsmen that stayed with us for the first three or four days." After Hurricane Sandy in October of 2012, the *New York Times* reported many businesses could not reopen due to lack of power. Losses from the storm were severe. The *Times* reported, "All told, the lost output from and overall effects of the storm could shave as much as 0.6 percentage points off annualized fourth-quarter economic growth, according to an analysis by IHS Global Insight." Beach towns along the New Jersey shore depended on sand removal and power for their businesses to recover, and in the long term, depended on constructing higher sand dunes to protect businesses and private houses.

Because all interdependencies cannot be foreseen, emergency management and business continuity planners need to incorporate planning tools into their planning processes that provide a process used to identify interdependencies. After doing this, planners can then meet with counterparts in the private and public sectors to ensure that their plans are congruent and that no holes exist, such as ensuring that when buses are available, so are the drivers to operate them. To facilitate the process of identifying interdependencies, we provide a preliminary methodology. We will present an example from the perspective of the public sector, but a similar example could be presented from the private sector perspective. For the public sector case, this methodology joins the four emergency management planning factors mentioned earlier—mitigation, preparedness, response, and recovery—with four general "content-area" factors: policies, people, processes, and suppliers, which are given in Figure 7.1, along with summary definitions. These are for illustrative purposes and in specific situations other "content" factors may be developed and substituted. As discussed elsewhere, different phases can be identified as well.

Given these factors for both the public and the private sectors, a methodology can be used to examine possible private–public interdependencies that occur in the four phases of the planning process. A grid that illustrates this is presented in Figure 7.2.

During the planning process regarding a specific crisis event, for each emergency management phase, planners can use the factors to identify potential interdependencies. These include intrasector and intersector interdependencies. To facilitate this process, several questions may be posed and answered:

- What does success mean for this phase of the process?
- What must "go right" for success to be achieved?
- What constraints exist that stand between the current situation and success?
- What interdependencies exist?
- What are some "action items" that need to be addressed, moving forward in the planning process?

| Definition of factors | | |
| --- | --- | --- |
| Note: The definitions below are to be taken in the context of a crisis event—from preparedness to full recovery. | | |
| Factor | Public sector | Private sector |
| Policies | Includes laws, regulations, and rules that govern how organizations communicate and operate. For example, the governor of a state mobilizes the National Guard. | Includes corporate procedures and rules that govern how organizations communicate and operate. For example, even in emergencies in banking dual control on transaction processing is required. |
| People | Includes employees of the public sector (police, fire, teachers, administrators, first responders) and citizens who draw on the services of the public and private sector. | Includes employees of the business organization and its customers. |
| Processes | Includes the processes used to deliver the services for the specific entity in question—e.g., for police protection; fire response; initiating governmental action. | Includes all of the processes used to fulfill orders from an organization's customers. This includes direct processes such as production and logistics, and indirect processes such as those governing the business continuity plan. |
| Suppliers | Includes the individuals, organizations, and infrastructure not directly employed or owned by the governmental entity, who provide products or services used by the entity to deliver its services to the public. Examples include third-party telecom providers, outsourcing companies, health care agencies, NGOs, businesses, and other governmental entities. | Includes the individuals, organizations, and infrastructure not directly employed or owned by the business or private organization who provide products or services used by the entity to deliver its services to its customers. Examples include third-party telecom providers, outsourcing companies, health care agencies, governmental agencies, and other businesses. |

**FIGURE 7.1**
Content-area factors.

| Interdependency grid | | | | | |
| --- | --- | --- | --- | --- | --- |
| | | Factors | | | |
| | Phase | Policies | People | Processes | Suppliers |
| Public sector | Mitigation | | | | |
| | Preparedness | | | | |
| | Response | | | | |
| | Recovery | | | | |

**FIGURE 7.2**
Private–public interdependencies.

For example, consider a coastal vacation town developing an emergency management plan for a hurricane and the mitigation phase of the process. In this phase, success might be identified as a reduction in damage from a past event using some metric, for example, a 90% reduction in damage to houses and businesses as measured in dollars, or, as discussed later, to successfully survive a 100-year storm. In the short term, we cannot control the force of the storm itself nor the direction. The necessary factor is that given the force of the storm, buffering mechanisms must exist to reduce the damage. There are many constraints including the existing configuration of beaches and dunes, houses and businesses, financial constraints, and human constraints including issues regarding evacuation and a desire for many people (residents and visitors) for ocean access and ocean views. Interdependencies are many and will be discussed next. Finally, given the interdependencies that do exist, action items would include strengthening relations for known interdependencies and coordinating crises event plans, and where interdependencies were unrecognized, initiating contact and following up with appropriate measures.

To illustrate how Figure 7.1 can be used to highlight interdependencies, consider mitigation and, for that phase, the issue of sand dunes. The following is illustrative only and does not go into the detail that would be required in a full analysis. Suppose that current policy and practice is that sand dunes should be at least 3.6 meters high (12 feet), although over time this has not been enforced. To mitigate damage from future storms, suppose, upon the recommendation of experts, the town is considering implementing a mandatory dune height to 6.7 meters (22 feet). This number would be the result of an analysis that defined success as the ability of the town to withstand a 100-year storm. Given these values, the Mitigation–Policy cell would identify the policies all of the levels of government that the town's policy would need to meet. The interdependencies might include various state and federal regulations, and policies of private insurance companies that might affect the policy decision. Other parties might include architects and engineering firms involved in the design of the mitigation step.

The Mitigation–People cell would involve homeowners of beachfront property and others whose property value might be affected by the dunes, and possible vacationers who might rent those houses during the summer beach season. In addition, local public employees such as police and fire departments may be involved as the process evolves. Interdependencies might include lawyers who might sue, protesting raising the dune height; private firms and business, such as restaurants, stores, and realtors, who might be affected by the changes; and surrounding towns whose citizens might also be impacted by the dunes and also by sand dredging operations to form them.

The Mitigation–Process cell would involve the actual process of claiming land and implementing the reconstruction effort. Here the United States Army Corps of Engineers may be involved as managing the effort, and other entities would be agencies of the local, state, and federal government including those concerned with public health. Areas where these agencies may be involved include funding, ensuring that any environmental or ecological impact falls within guidelines and satisfies existing regulations, and identifying all of the private companies that will be involved in the construction effort.

The Mitigation–Suppliers effort would focus on the organizations and infrastructure necessary to implement the planned construction. This would include ensuring availability of equipment, personnel, and the timing of construction such that it meets the proposed schedule. Here, suppliers might include private companies that use equipment that takes sand from the ocean floor and uses it for dune construction. Given that efforts to resist the plan may arise, other supply partners might include public relations firms hired to convince the public of the necessity for the mitigation plans.

This is a hypothetical example for just one aspect of one phase of the emergency management process but consider all of the possible interdependencies already identified:

- State and federal agencies tasked with standards for beach construction
- Local citizens
- Vacationers
- Insurance companies
- Architects
- Engineering firms
- Police and fire departments
- Lawyers
- Local businesses
- Surrounding communities
- U.S. Army Corps of Engineers
- Federal and state agencies tasked with environmental and ecological protection
- Funding agencies
- Hospitals and health agencies
- Shoreline engineering companies
- Construction firms
- Labor unions
- Public relations firms

Similar steps would occur for the preparedness, response, and recovery phases. Here, given the hypothetical nature of the discussion, the identities are general. In a real analysis, specific businesses may be involved and when done analyzing all four phases, hundreds of interdependencies between the town and various other public and private entities would be identified. For example, in the response–supplier discussion, providers of response services would be discussed, such as companies that might supply trucks and bulldozers to clear sand blown by hurricane-force winds and flooding off streets. Other interdependencies may include companies with their own business continuity plans, such as a telecommunications provider who may have wires down, thousands of poles to rebuild, and power to restore to customers. In this case, the business continuity plan for this company would be integrated with the emergence of a response plan for the town.

For the private sector, the definition of phases of a company's business continuity plan might differ. For example, the phases might be defined to include business impact analysis, risk assessment, strategy development, emergency response, and full recovery. However the specific phases are defined, a similar process can take place. The key point is that going through the structured process will identify interdependencies that will mutually benefit parties in both the private and public sectors.

## Risk Analysis Framework and Assessment for Disaster Risks

Urban development, modern human settlements, production, transportation, and consumption have contributed to emerging natural disasters and are associated with shifting sociocultural and technological paradigms. Natural disasters' risks are specific in their characteristics and typically of "low frequency but of catastrophic consequences" in terms of human life and health, economic loss, and associated environmental damage (Ikeda et al., 2006).

The evolution of "risk" traveled from a conventional view of it as "an expected value of the probability of a hazardous event occurring times the magnitude of the consequence of the hazard" to an ontological or sociological concept that provides the tools for consideration of a wide range of sociocultural characteristics of disaster risks. From the attempts to establish an interdisciplinary concept of risk, the analysis leads to a straightforward conjecture that can be understood as "a potential for the realization of unwanted, adverse consequences to human life, health, property, or the environment" (Ikeda et al., 2006).

An effect is a deviation from the expected—positive and/or negative—and uncertainty is the state, even partial, of deficiency of information related to, understanding, or knowledge of an event, its consequence, or likelihood (Pojasek, 2013a).

A strategic plan for climate change is founded on the assessment of the likelihood of an event and its consequences for systems in the megacity's planning areas. Megacities must identify and estimate consequences (economic, ecological, social, cultural, and legal) of a particular climate change impact and assess these consequences and factor in the estimated scale of the impact, such as population density and the geographical area of the impact affected by a projected climate change impact. Another thing to consider when assessing risk is the cumulative costs associated with a higher frequency of minor events (The Climate Impacts Group et al., 2007).

Kaplan and Garrick (1981) proposed the following expression of the risk concept as a risk triplet. The risk triplet consists of scenario (S), likelihood (P), and possible consequence (D) in relation to three basic questions of risk analysis:

1. What is the nature of disaster events that can occur?
2. How likely is a particular event?
3. What are the consequences?

The triplet expression of risk: Risk = R{<Si, Pi, Di>},(i = 1, 2, ...)

1. Si: What will happen?
2. Pi: How often?
3. Di: What consequences? (Kaplan and Garrick, 1981)

Scenarios Pi and Di embody the scientific knowledge of events. Pi represents a set of likelihoods concerning event frequency, probability, or ambiguity. Di is a set of consequences concerning potential damage to humans, animals, plants, and the environment. Nevertheless, Si can incorporate the subsequent questions and conditions of factors related to ontological and sociocultural dynamics in relation to anthropocentric activities (Ikeda et al., 2006).

Carolan (2004) stated that if ontology is multiple, then it is also political. Environmental crises can neither easily be seen nor quickly controlled. What the world faces today is likely to become more complex and increasingly multiple as those issues distance themselves further from the epistemological scope. Sociocultural values and collective perceptions of climate risk can be the consequence of defied resistance and challenges among different communities, shaped by cultural values and societal patterns. Despite these complex issues, multiplicity does not necessarily mean fragmentation (Carolan, 2004). Sociocultural perspectives seek to understand human behavior as well as how opinions of the general public can compel or constrain political, economic, and social action to address particular risks. To a great extent, "whoever controls the definition of dangerous climate change controls the rational solution to the problem. If danger is defined one way, then one set of solutions will emerge as the most cost effective or the safest" (Leiserowitz, 2005). These ontological and sociocultural dynamics are relevant factors to the collective-constructed perception of climate change risks. Risk is an objective hazard, threat, or danger that is foreseeably facilitated through social and cultural factors, and these dynamics influence judgments, perceptions, and behaviors.

While conducting a risk analysis, the most critical endpoints need to be identified and assessed for possible impacts to megacities' dwellers, their communities, and environment to facilitate a risk management approach that provides guidance on how to proactively respond to risks. Endpoint indices (or endpoint interpretations) can be inferred as "outcomes vulnerability." This concept considers vulnerability as the (potential) net impact of climate change on a specific exposure unit (which can be biophysical or social) after feasible adaptations are taken into account (Fellmann, 2012).

Assessing endpoints and indicating the particular nature of each risk is important for determining the likelihood and the degree of damage. To understand the scale and degree of damage, climate specialists look at the loss of environmental assets, environmental quality, lives, and infrastructure. They also try to determine if the impact of the event is within reach of management to cope with outcomes that pose a potential threat to human ethics or morality. These external impacts analysis help to clarify scenarios Si in the problem formulation stage at the scientific or objective evaluation of Pi and Di, because we may need other assessment schemes or tools tailored to the nature of each possible risk scenario. Risk analysis guides the process of understanding these scenarios either qualitatively or quantitatively, applying the risk triplet R = R {Si, Pi, Di} based on an analytical framework of risk analysis (Ikeda et al., 2006) and building the mechanisms that can be part of a comprehensive integrated framework of risk management and adaptation for megacities.

## Conclusions and Future Challenges

While planning for disasters in both the public and private sectors is drawing more attention and while the quality of the planning processes and the plans themselves is improving, a challenge remains to recognize the interdependencies between private and public sector plans and to develop a methodology to effectively manage them. Postmortem analyses of many disaster events have indicated that individual entities often do well within the limited scope of their expertise, but when systemic performance is assessed, more holes exist. This was true for Hurricane Katrina where the Coast Guard performed admirably and some businesses plans effectively supported recovery efforts. The greater challenge existed when hidden policy and resource allocation and coordination issues surfaced.

Perhaps the most important step in recognizing these interdependencies is ongoing communication among all relevant parties. This not only includes normal ongoing communications and meeting at professional gatherings, but more substantively sharing plans and even participating in drills to exercise those plans. A second step, more as part of the planning process itself, is to implement a structured methodology aimed at triggering ideas and identifying issues. Grids with other definitions are certainly viable alternatives as are other completely different methods. Forward-looking organizations create infrastructures resilient enough to eliminate or reduce the impact of disasters. The important point is that moving forward, emergency management planners from the public sector and business continuity planners from the private sector need to communicate and collaborate so they can develop and implement solutions that more effectively address future crisis events.

## List of Terms

**Plan–Do–Check–Act (PDCA):** An iterative four-step management method used in business for the control and continuous improvement of processes and products.

**Risk triplet:** A set of three questions used to identify the quantitative measurement of risk in a given scenario.

# 8

## Building Resilience and Sustainable Development in the Face of Disasters and Climate Change

### Benefits of Building Resilience: British Standard 65000—Organizational Resilience

Organizational resilience management is a management framework comprised of planning, decision making, and execution needed to anticipate, prevent (if conditions allow it), prepare, and respond to disruptive events and changes in the environment that detrimentally affect organizations (Leflar and Siegel, 2013). In this case, the organization "megacity" benefits from British Standard 65000:2014 (BS 65000:2014) as it strengthens its capacity to prevent, minimize, manage, and tolerate risks from climate change impacts.

BS 65000:2014 provides guidance on building organizational resilience by identifying the core requirements that a megacity must meet to be considered resilient. The standard also helps city leaders identify and recommend practices that are already established in other standards and disciplines. What this means in practice is that BS 65000:2014 defines value as a function of risk and return. Thus, city leaders' decision-making processes and initiatives can preserve or increase value, but they can also gradually erode value. Despite the wishful thinking of many city leaders, risk is unavoidable. It is a central part of the quest of value. As adapting to climate change becomes fundamentally important to the resilience of the population, strategic planning leaders need to think of risk as not something to avoid. Instead, they should view it as something to manage across all sectors of their city's departments and divisions that provide public safety, health care services, communications, transportation, and security.

The BS 65000 framework provides a management system that forms the basis for establishing and building resilience. It highlights the need for systems and processes that will help spur good decision-making. The framework enables a city to prepare, initiate, and manage a climate change adaptation program. The standard provides guidance on enhancing organizational resilience and articulating the crucial benefits for adaptation plans.

Existing standards within the crisis and business continuity management domains, which have an impact on the overall governance of an organization, enhance their practices by integrating the disciplines coordinating the necessary steps and initiatives for a city's resilience (BS 65000:2014, 2014).

Implementing BS 65000 solutions will promote resiliency for megacities, enabling them to adapt effectively to unexpected and disruptive events. BS 65000 can help reduce costs and maximize operational efficiency by minimizing hazards that exacerbate vulnerabilities. Implementing these solutions is crucial to the survival of the megacity because resilient organizations prosper regardless of the adverse climatic conditions and its impacts.

## Building Capacity

Managing and minimizing risks provides capital investment opportunities for megacities through initiatives undertaken by local governments and with the involvement of the private sector (see Chapter 6). Capital investment must focus on upgrading the megacity's infrastructure through deep energy retrofits by involving owners and investors in carrying them out across the city's infrastructure—supported by a strategic-minded finance community with a focus on the fundamental facilities and systems serving the city, including the services and facilities necessary for its economic function.

Adaptation plans cannot be the sole strategy for megacities to cope with climate change impacts. Efficient functioning of the megacity depends on its ability to reach a desired level of resilience. Adaptation and disaster risk reduction (DRR) strategies need to be all-inclusive and complement each other, focusing on overall risks, development conditions, and local area performance (Brugmann, 2012). From an environmental science approach, where climate change adaptation is emerging, adaptation strategies translate onto shifting environmental conditions. DRR strategies can extend to social, physical, and economic factors, and complement adaptation strategies. DRR strategies are able to identify the wider constraints that determine vulnerability while adaptation strategies focus on developing hazard forecasting and early warning systems. Theoretically, DRR extends beyond disaster preparedness measures exclusively (Venton and La Trobe, 2008).

## Planning for Resilience in a Vulnerable World

To identify megacities' vulnerabilities, leaders need to frame strategies that have clear objectives. These approaches build upon regulations and compliance that direct away from and lessen the risks that increase vulnerabilities. The prevention and alleviation of those threats become part of an agenda that emphasizes measures of outcomes and community sustainability. As the number of urban settlers increases rapidly around the world, the social aspect of this plan becomes fundamental to progressive development because infrastructure is a matter of interest and importance for community resilience.

Risk reduction strategies must lead to actionable concepts and practices toward minimizing threats as well as evaluating and managing the issues that lead to disasters. The reduction of risks is the fundamental concept of DRR, which includes reducing exposure to hazards, addressing the megacities' vulnerability, effective environmental management (poor environmental stewardship is one of the root causes of disaster risk), and a proactive approach toward severe weather events and climate variability.

Several factors help leaders of megacities to build resilience. Action through policies and strategic planning should integrate the risks associated with climatic events, public health, and the megacity's infrastructure. Conventional planning processes, project design, and development decision-making need to become the dynamic foundation for disaster risk reduction. Additionally, megacities' leaders must consider building capacity for local institutions to move forward large, complex plans and to match local demand for resilience with financial support (Brugmann, 2012). Organizational resilience forces leaders into building the capacity to adapt (Denhardt and Denhardt, 2009) and involves strategic problem-solving to create sustainable change for the organization and the environment. Building resilience also means adapting quickly to circumstances and conditions outside the standard model (Hollnagel et al., 2006). Rapid urbanization causing environmental degradation amplifies vulnerability and exposure to megacities. Therefore, exposure needs to be minimized, vulnerability reduced, and resilience capacity built up to reduce disaster and climate change risk. Risk reduction plans for both disasters and climate change need to complement each other and entail a continual effort across economic, social, cultural, environmental, institutional, and political spheres to move from vulnerability to resilience (Turnbull et al., 2013).

## Continuity Plans and Resilience

Organizational structure and the extent of community participation are significant considerations for megacities' plans to become resilient. Efforts here can assist local governments in working along with their communities, addressing vulnerabilities, and solidifying the relationships. A predictor of the megacity's organizational survival and resilience is its readiness to implement a predisaster and business continuity strategic plan (Cox Downey, 2015), and consequently mitigate the risks associated with climate change.

## Governance of Megacities: Exploring How Governance Matters to Resilience

Governance is concerned with improving the performance of organizations for the benefit of shareholders, stakeholders, and economic growth. It focuses on the conduct of, and relationships between, the board of directors, managers, and shareholders. Megacity leaders can benefit from a comprehensive framework that addresses good governance principles to support and improve the already established program the city has implemented. Improving the performance of a city's operations must be a priority for leaders who should focus on the conduct of, and relationships between, the board of directors, managers, and shareholders. The Organisation for Economic Co-operation and Development's (OECD) definition of corporate governance refers to the processes by which organizations are directed, controlled, and held accountable. It encompasses authority, accountability, stewardship, leadership, direction, and control exercised in the organization (OECD, 2004).

Effective governance must be assured through strong commitment by people at all levels within the entity. A policy that supports the challenges and demands of governing must be in place. With a clear policy statement that encompasses the entity's commitment to good governance, cities will be able to cover all of the necessary components of good corporate governance. The governing body must understand, promote, and be responsible for good governance.

## Adopting a Philosophy and Strategy of Continuous Improvement in Governance Performance

The OECD has stated that there is no single model of corporate governance, that is, no one-size-fits-all model. Pluralism, flexibility, and adaptability are required in corporate governance. These requirements are true for cities, too, because they must understand there is no common denominator to good governance. Instead, a number of common elements serve as the foundation for good corporate governance (Australian Standard [AS] 8000-2003, 2003). Risk is an important element of corporate governance. The concept of a control environment has to start from the top of an agency. Risk management establishes a process of identifying, analyzing, and treating risks that could prevent any nation's government agencies from effectively achieving their business objectives (Barrett, 2002). It includes establishing links between risks/returns and resourcing priorities. Risk management includes putting appropriate control structures in place to manage risk throughout an organization by developing sound risk management plans. These plans cover activities as diverse as reviews of operating performance, best uses of information technology, increased competition, outsourcing, performance management and information, professional development, and staff appraisal. These activities also include client surveys, reconciliations of accounts, setting and adhering to appropriate delegations, and proper segregation of duties (Barrett, 1999).

Mayors should follow the recommendations and guidance of the Australian Standard (AS 8000) that defines governance and would work for a megacity. Accountability and responsibility play a key role in governance and the willingness of the owners of the entity to behave like owners and put the right of ownership into effect. These actions cause leaders to express their views to boards of directors or city leaders. They can also organize and exercise their shareholder franchise if they do not receive a satisfactory response. The shareholders as owners elect the directors to run the entity on their behalf and hold them accountable for the results. It is for the shareholders to call the directors to account if they appear to be failing in their stewardship. The rights and obligations of the shareholders expressed in AS 8000 can be followed by city leaders to emphasize effective governance and for the development and implementation of a risk management plan that can be tailored to their current strategy.

## Rights and Obligations of Shareholders

AS 8000 states that shareholders should

- Remain as informed as possible about the activities of the entity.
- See themselves as owners, not just investors; their responsibility as shareholders increases with the size of their shareholding.
- Make a sufficient analysis in order to vote in an informed manner on all issues raised at general meetings. When appropriate, reasons for voting against a motion should be made known to the board beforehand.
- Avoid involvement in the entity's day-to-day operations.
- Have an informed interest in the composition of boards of directors with particular reference to concentrations of decision-making power.
- Appoint a core group of independent directors of appropriate caliber and experience.
- Have an informed interest in the structure of boards and, in particular, the appointment of appropriate committees of the board—especially the auditing committee.

## Governance for Health

The concept of "governance for health" can best be illustrated as the culmination of three waves in the expansion of health policy, from intersectoral action, to healthy public policy, to the "health in all policies" approach, all of which are now integrated in whole-of-government and whole-of-society approaches to health and well-being.

**Professor Ilona Kickbusch**

## Improving Megacity Services and Quality of Life

The Brundtland commission's report conceptualizes the significance of sustainable development (Visser and Brundtland, 2009):

> Sustainable Development seeks to meet the needs and aspirations of the present without compromising the ability to meet those of the future. Far from requiring the cessation of economic growth, it recognizes that the problems of poverty and underdevelopment cannot be solved unless we have a new era of growth in which developing countries play a large role and reap large benefits.

Some analysts refer to megacities as dystopian places where unequal distribution of economic growth and increasing population density leads to inequality, chaos, and political and social instability, particularly in the developing world. These mega-centers are seen as "feral cities" in which conflict will be "crowded, connected and coastal" (Evans, 2015). This makes it imperative to create and expand developing policies and programs at a large scale to advance sustainability of buildings, businesses, and urban forms, ensuring nature's ability to function overtime. Promoting efficiency and developing local resources to strengthen the local economy and supporting the development of the economic, human, social, and natural capital needs to be part of the megacity's leader's agenda. These actions will enable expanding public open space systems and advancing the programs to improve efficiency in buildings, business, and urban forms to create the right environment for sustainable growth (Centre for Sustainable Community Development, 2014).

## Achieving Sustainable Development

The intent of methodologies is not to find solutions but rather to provide guidelines to facilitate the path to find such solutions. Nonetheless, defining methodologies to set up indicators to measure and direct megacities' performance is necessary to deliver the required services for the urban conglomerate to operate efficiently and provide quality of life for its residents. Megacities' leaders need to build strategic frameworks to deliver a clear defined performance and the ability to measure progress for improvements. ISO 37120:2014 is a standard that can help megacities with an interconnected and all-inclusive plan that can be quantified in a comparable and demonstrable method to achieve sustainable development.

Accelerated population growth in megacities is leading to deterioration in the quality of life. Urban citizens are faced with many inequalities that affect their well-being and put pressure on the city's budget, health care system, resources, and operational efficiency. Measuring progress over time and providing comparative lessons in a local and global context from other megacities' performance is key for continuous improvement. ISO 37120 helps guide policy, planning, and management, integrating stakeholders and other sectors of society. It supports the case of establishing policies to promote the well-being of its inhabitants with the enclosure of a comprehensive and economically viable framework to create sustainable growth in order to assist megacity leaders in developing and implementing practical actions toward resilience. The following shows the highlighted ISO 37120 benefits of standardized indicators:

- More effective governance and delivery of services
- Local and international benchmarking and planning
- Informed decision making for policy makers and city managers
- Learning across cities
- Recognition by international entities
- Leverage for funding by cities with senior levels of government
- Framework for sustainability planning
- Transparency and open data for investment attractiveness (World Council on City Data, 2014)

## Disaster Risk Management and Sustainable Development for Public Health

Building resilience for disasters requires a strong sense of urgency in the context of sustainable development and integration into present and future strategic planning and budgets. With that in mind, disaster risk management becomes the main driver of sustainable development by strengthening risk assessment, disaster prevention, and humanitarian responses. Communication and good relations with individuals and communities are important to reducing risks because it is a long-term development process. Health care systems provide core capacities for disaster risk management for health. In that capacity, some megacities in developing countries do not have the resources and infrastructure to confront such challenges. In contrast, megacities in developed countries are often more resilient and better prepared for disasters and, therefore, more able to cope with climate change. Primary health care (PHC) delivers basic services plus the response to emergencies, along with policies and strategies focusing on PHC help minimize vulnerability and prepare as well as adapt megacities for disasters (World Health Organization, 2011).

After disaster strikes, attention is directed to care for acute health conditions and specialist interventions. It is important to note that typically chronic and preexisting conditions attest the largest burden of disease.

Understanding these facts can well-prepare and organize a community, and reduce risks and the impact of disasters. In health care emergency response, the first hours after the disaster are vital to saving lives before external help arrives (World Health Organization, 2011).

## Biophilic Megacity

The term *biophilia* was coined by American biologist, researcher, theorist, naturalist, and author E. O. Wilson to describe the tendency for humans to have an innately emotional response to other living organisms (Wilson, 1984). Human health and the well-being of a population are essential elements for a sustainable, ecological balance; the link between those two elements and nature have been studied and understood by the health and psychology fields. From that point, there is increased interest among researchers in the relationship of the impacts of nature in urban environments on the human condition (Reeve et al., 2012).

Increasing the use of natural ecosystems to address biodiversity, landscape, food production, and water management contributes significantly to elements in the built environment because they are all interconnected to a resilient megacity. The challenges of declining biodiversity in the face of global urbanization is obvious in the majority of megacities that are stressed by the increasing number of urban settlers and the lack of resources (financial, social services, etc.) to accommodate the new dwellers. Then, there is a question that presents a new paradigm: How can megacities improve their urban environment and address biodiversity challenges? First, leaders need to work with stakeholders from the beginning. There is not a fixed methodology to improve urban environments, because each megacity's geographic location, and infrastructure and building design have advantages and limitations for improvement. Therefore, key local stakeholders (governmental and nongovernmental institutions, intermediary institutions, industry representatives, community engagement groups, and so on) play an important role in finding the impediments to enhance the megacity's built environment, and considering specific social and economic contexts of the megacity and the state of their current environmental situation. Implementing such initiatives requires a bottom-up approach with the collaboration of all stakeholders addressing urban issues.

In regard to biodiversity challenges faced by megacities, biophilia can address these issues with the implementation of large-scale landscaping that offers open space into an important biodiversity corridor. Urban farming has many benefits to the environment and provides mitigation by increasing food security, landscape multifunctionality, biodiversity habitats, and carbon sequestration capacity. Biosequestration plays a key role as it removes "legacy" emissions from the atmosphere in addition to improvements in storm-water management through rainfall retention and consequently better use of water supply (Bilsborough, 2015). Climate change puts stressors onto the basic services and resources the megacity's population require, and these risks unleash the potential for nature to address other challenges related to climate change mitigation and adaptation. In a changing climate, the functionality provided by urban green space becomes increasingly important and a path to a "biophilic megacity." The use of urban green space offers significant potential in moderating the increase in summer temperatures expected with climate change (Gill et al., 2007). The relationship between sustainable development, biophilia, and urban design became an important topic for researchers and city leaders with a changing climate. Urban environments may provide certain environmental benefits for megacities as it contains a variety of ecological and green assets from parks to trees to rivers and riparian habitats (Beatley and Newman, 2013). Enhancing green elements and features of these living and work environments offers a viable alternative for megacities and urban environments to sustain health and well-being for the urban population. Beatley and Newman (2013) presented examples of biophilic design interventions (see Table 8.1).

To propel innovation it is essential to highlight the importance of management, government leadership supporting policy and adaptive measures, and building design standards that provide incentives for investing. Political cycles play a role in decision-making; therefore, actions that go beyond traditional economies and short-term goals must be implemented at a top level of governance throughout all levels of the megacity's government. The inclusion of biophilic urban elements is not present in existing regulations and

**TABLE 8.1**

Biophilic City Design Elements across Scales

| Scale | Biophilic Design Elements |
|---|---|
| Building | Green rooftops, sky gardens and green atria, rooftop garden, green walls, daylight interior spaces, green courtyards |
| Block | Clustered housing around green areas, native species yards and spaces, green streets, urban trees |
| Street | Low impact development (LID), vegetated swales and skinny streets, edible landscaping, high degree of permeability, neighborhood stream daylighting and restoration |
| Neighborhood | Urban forests, ecology parks, community gardens, neighborhood parks/pocket parks, greening greyfields and brownfields, community urban creeks and riparian areas, urban ecological networks, green schools, city tree canopy |
| Community | Community forest/community orchards, greening utility corridors, region river systems/floodplains, riparian systems |
| Region | Regional green space systems, greening major transport corridors |

*Source:* Girling, C., and Kellett, R., 2005, *Skinny Streets and Green Neighborhoods: Design for Environment and Community*, Washington, DC: Island Press. First appeared in Beatley, T., 2010, *Biophilic Cities: Integrating Nature into Urban Design and Planning*, Washington, DC: Island Press.

planning requirements. This lack needs to be addressed by government and private sector leaders because it is fundamental to sustainable management of megacities.

There is an alternative frame of reference, coined by Israeli-American sociologist Aaron Antonovsky, called *salutogenesis*. It refers to a new approach to health promotion and needs assessment. Salutogenesis seeks the promotion of good health as the primary goal rather than the curing of disease (or even the preventing of disease). Salutogenic orientation is opposed to the pathogenic orientation underpinning the biomedical and National Health Service (United Kingdom) worlds, which focus on the particular diagnosed disease entity. This approach examines the underlying social constructs, the broader picture, in order to both define the health problem and to search for coping resources or mechanisms (Royal College of Midwives, 2002).

There is an understanding within management about the beneficial health effects of including biophilia elements in urban environments and intertwining this approach with any efforts toward a sustainable management strategy and the improvement of urban environments and public health. As climate change builds stressors for public health, Antonovsky's approach to health presents a critical look to health challenges and offers a viable alternative to the limitations of risk factor approaches for conceptualizing and conducting research on health (Antonovsky, 1996).

## Thinking More Broadly about Urban Resilience

*Joe Leitmann*

The global community has recently passed the urban tipping point with over half of the world's population now living in cities. Most of these 3.3 billion people reside in developing countries where the vast majority of future urban growth is expected. This event has implications for the world's poor because urbanization is highly correlated with reductions in poverty and inequality and improvements in access to essential services (World Bank, 2008). As highlighted in the State of the World's Cities report, "urbanization has become a positive force for transformation that makes countries more advanced, developed and richer, in most cases" (United Nations Human Settlement Programme, 2013). Cities are engines of economic growth that account for more than 80% of global GDP and can drive a nation's prosperity as well as an individual's access to opportunity (Dobbs, 2011).

But urbanization also brings challenges. As people and their assets increasingly concentrate in cities, they become highly dependent on infrastructure networks, communications systems, supply chains, and utility connections for their well-being. Natural and man-made disruptions or shocks to this complex organism can have a catastrophic impact on a city's ability to meet the most basic needs of its citizens and reverse decades of economic development gains (Benson and Clay, 2004). Moreover, rapid and unplanned urbanization is itself a driver of risk. Development in high-risk areas, such as hillside slopes, floodplains, or subsiding land, is often uncontrolled, with the poor and the vulnerable settling in hazardous areas because they are affordable (Jha et al., 2013). And, in many developing countries, cities are continuing to expand public and private infrastructure in vulnerable areas without proper assessment of risks. Which cities will grow, how fast, and which hazards they will confront are less predictable (World Bank, 2010b). In tandem with this, the increasing concentration of people, assets, and infrastructure in cities is amplifying their exposure to hazards of all sorts, but especially to natural hazards and the impacts of climate change.

These risks result in significant economic costs. The United Nation's Global Assessment Report on Disaster Risk Reduction highlights that economic losses from disasters are now reaching an average of $250 billion to $300 billion each year; future annual losses from earthquakes, cyclones, flooding, and tsunamis are expected to average $314 billion in the built environment alone (United Nations Office for Disaster Risk Reduction [UNISDR], 2015). Moreover, sea-level rise and subsidence in the 136 largest coastal cities could result in losses of $1 trillion or more per year by 2050 (Hallegatte et al., 2013). Indeed, cities' greatest strengths for economic growth—efficiency and interrelation of infrastructure and density of population—can also be their potential weaknesses to cascading failure from the impacts of disasters (Graham, 2010).

Climate-related shocks and stresses result in a broad spectrum of interdependent effects on people, infrastructure, and urban systems. In many ways, climate change is a threat multiplier. Sea-level rise, changing rainfall patterns and rising temperatures, more intense storms, and prolonged periods of drought set in motion cascading impacts with the potential to override system thresholds leading to catastrophic consequences (Rumbaitis del Rio and Sjögren, 2014). Recent events such as Hurricane Sandy and Typhoon Haiyan demonstrate the impact of extreme events on cities. For example, Typhoon Haiyan affected 16 million people, caused 6,000 deaths, displaced 4 million people, and destroyed or damaged 1.1 million homes (USAID, 2014). Although the effects were spread across 12,000 villages and 44 provinces in the Philippines, one-third of deaths occurred in Tacloban City because of the population concentration and substandard housing conditions (Rumbaitis del Rio and Sjögren, 2014).

Natural disasters, whether rapid or slow onset, are not the only risks facing cities (see Figure 8.1 for a classification of urban hazards). In both developed and developing countries, cities are vulnerable to economic downturns, social upheaval, dependence on single industries, public health epidemics, terrorism, or the failure of infrastructure due to obsolescence. These shocks can have devastating effects, bringing some or all of an urban system to a

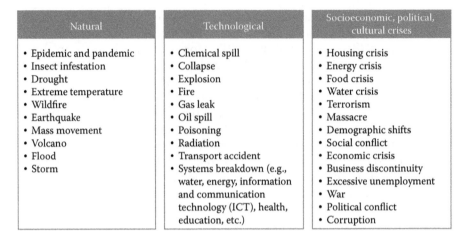

| Natural | Technological | Socioeconomic, political, cultural crises |
|---|---|---|
| • Epidemic and pandemic<br>• Insect infestation<br>• Drought<br>• Extreme temperature<br>• Wildfire<br>• Earthquake<br>• Mass movement<br>• Volcano<br>• Flood<br>• Storm | • Chemical spill<br>• Collapse<br>• Explosion<br>• Fire<br>• Gas leak<br>• Oil spill<br>• Poisoning<br>• Radiation<br>• Transport accident<br>• Systems breakdown (e.g., water, energy, information and communication technology (ICT), health, education, etc.) | • Housing crisis<br>• Energy crisis<br>• Food crisis<br>• Water crisis<br>• Terrorism<br>• Massacre<br>• Demographic shifts<br>• Social conflict<br>• Economic crisis<br>• Business discontinuity<br>• Excessive unemployment<br>• War<br>• Political conflict<br>• Corruption |

**FIGURE 8.1**
Classification of urban hazards (shocks and stresses). (Adapted from UN-Habitat's City Resilience Profiling Tool (CRPT), which is based on the classification of hazards by EM-DAT and PreventionWeb.)

halt, and possibly causing loss of life and asset damage. For example, the Arab Spring resulted in US$800 billion in lost output (HSBC, 2013) and over 50,000 deaths (Ibish, 2012; Tarnawski, 2013) in the seven hardest-hit countries. In addition, Singapore's exposure to SARS cost the government nearly US$570 million (Sitathan, 2003), and the collapse of the Rana Plaza building in Dhaka, Bangladesh, in April 2013 resulted in the deaths of over 1,000 people.

Not all shocks and stresses are equal, especially when man-made crises are included on the same menu as natural hazards. For example, Chileans expect and are prepared for earthquakes; they were far less prepared for the shock to the urban system caused by the Chilean Winter—the massive protests of August 2011 led by students demanding a new framework for education in the country. Tens of thousands of people took to the streets of Santiago, brought traffic to a halt day after day, and occupied and damaged public school buildings. While an earthquake is clearly a negative event, it is difficult to claim that public demonstrations for education reform have the same status, even if they result in damage to public property and local businesses. Indeed, it could even be argued that one benefit of urban agglomeration is to allow the population to have more voice and to give the government an incentive to be more accountable and reactive to the desires of its citizens, even if it creates more social perturbations and temporary losses. However, while the origins and long-term impacts of shocks may differ dramatically, the necessity of the city to absorb, adapt, and continue functioning in the short-term remains the same (World Bank, 2014).

There is also a distinction between acute shocks and chronic stresses. Shocks are sudden events that impact the vulnerability of a system. There are many different types of shocks that can strike at different levels, including disease outbreaks, floods, high winds, landslides, droughts, or earthquakes. Outbreaks of fighting or violence, or severe economic volatility, could be included in this category as well. Stresses are longer-term trends that undermine the performance of a given system and increase the vulnerability of actors within it. These can include natural resource degradation, loss of agricultural production, demographic changes (e.g., aging and depopulation), climate change, political instability, or economic decline (Department for International Development [DfID], 2011). A significant stress facing many cities in developing countries is urbanization itself due to the pressure it places on urban systems and the delivery of essential services. Stresses can be cumulative, compounding gradually until a tipping point is reached and transformed into a shock.

Urban resilience is not limited to the administrative boundaries of the city. Broader regional dynamics inform and affect city resilience. These dynamics play out between cities and surrounding rural areas, other cities and, at times, other countries. They are informed by a variety of interdependent engagements. They can reinforce a city's resilience or undermine it, depending on the circumstances and specific event. For example, the 2013 summer drought in the western United States caused a wildfire in Yosemite National Park affecting power and water supplies to San Francisco, about 200 miles to the west. Hydroelectric power was interrupted, forcing the city to seek power elsewhere at an additional cost of $600,000. The globalization of markets and heightened reliance on global supply chains means that risk in one city or region can translate into impacts in another city or region. This became quite evident in 2011 for hard-drive suppliers after the floods in Thailand, and for Toyota after the earthquake and tsunami in Japan. Understanding these linkages, flashpoints, supply-chain chokepoints, and areas for cooperation are essential to developing resilient cities (World Bank, 2014).

Shocks have an impact on all aspects of development. Impacts are felt directly through the loss of lives, livelihoods, and infrastructure, and indirectly through the diversion of funds from development to emergency relief and reconstruction (DfID, 2005). Moreover, shocks disproportionately affect the urban poor. For example, during the 2011 Thailand floods, 73% of low-income households in Bangkok were affected compared to only 21% of the total city population (UNISDR, 2015). Moreover, a country that experienced significant violence over the period 1981 to 2005 has an extreme poverty rate 21 percentage points higher than a country with no violence (World Bank, 2011).

Informal settlements in urban areas are particularly vulnerable. Climate change poses a threat to urban residents and at the same time is expected to drive further urbanization, due to rural-urban migration, ultimately placing more people at risk. Migrants often settle in slums that are especially vulnerable to a wide range of risks, including those stemming from climate change and natural hazards. The circumstances of migrants are mainly due to the location of slums on at-risk land, overcrowded living conditions, the lack of adequate infrastructure and services, unsafe housing, inadequate nutrition, and poor health (Baker, 2012b).

Acute shocks and chronic stresses can also have a profound and lasting impact on human development. Disaster losses are often linked with or exacerbated by poverty and vulnerabilities of the poor that stem from socioeconomic and environmental imbalances. The losses are not limited to natural disasters. The 2008 financial crisis contributed to many forms of poverty in Eastern Europe. In Vilnius, Lithuania, for example, the number of applicants to get support from food banks increased 6.5 times in one year; and in Tallin, Estonia, the number of people receiving subsistence benefits increased by 63% over the same period (URBACT, 2010). In turn, shocks may push people into poverty and affect the ability of the poor to rise out of poverty because constant exposure to such threats and income shocks leads them to stay poor and vulnerable, setting in motion a vicious cycle (UNISDR, 2008). Without more resilient development, climate change could force more than 100 million people into extreme poverty by 2030 (Hallegatte et al., 2016).

The development community has a window of opportunity to make cities and the urban poor more resilient. Sustainable Development Goal No. 11 calls on the world to "make cities inclusive, safe, resilient and sustainable." COP 21 highlighted the key role that urban areas play, not only in mitigating emissions but also in adapting to a changing climate. The Habitat III Conference (October 2016, Quito) represents an opportunity to influence the New Urban Agenda by underscoring the importance of resilience for the socioeconomic and spatial development of cities and the well-being of their citizens, especially the urban poor (United Nations Human Settlement Programme, 2015).

---

## List of Terms

**Biophilia:** The tendency for humans to have an innately emotional response to other living organisms.

# 9

## Resilient Communities through Environmental Planning and Design: Accounting for Future Scenarios and Regenerative Capacity

Wendi Goldsmith

As we face global trends of climate change, urbanization of world populations, loss of habitats and ecological connectivity, and rising impacts to people and property due to natural disasters, those responsible for design of the built environment can potentially shape the well-being of future generations in terms of public health (or risk) and economic opportunity (or burden). Indeed, one of the most important tools for human adaptation is the planning and engineering of environmentally sound solutions, incorporating information gained incrementally to anticipate and holistically address future and ongoing needs. Sustainable development has been recognized for decades as the aspirational goal to guide universal action.

The definitions of sustainability, though numerous, share in depicting human populations and their economic systems nested within the finite resource system of planet Earth. The triple bottom line of sustainability accounts for social, economic, and environmental parameters (Fiksel et al., 2014). These ideas are not new, yet they remain outside the normal choices of most people and organizations. Even as we gain an increasingly clear understanding that conditions are changing, and the rate of change is increasing, uptake of new approaches to address root causes remains slow.

This chapter provides a roadmap for integration among professionals across relevant disciplines. What seemed to some like distant or hypothetical problems in modern society have become real as Central Europe experiences high mortality during heatwaves, California experiences drought levels that jeopardize public and agricultural water supplies, thousands die in the Gulf Coast during Hurricane Katrina, and dozens die in urban greater New York City during Superstorm Sandy. Many disasters cause breakdowns in critical infrastructure (power, water, sewer, and communications), first responder capacity (roads, bridges, fuel, and supplies), and medical operations (loss of staff accessibility, backup power system failures, and supply shortages). Other impacts are less direct, such as untreated wastewater discharges and stirred-up industrial pollution spread by flooding through neighborhoods and habitats. In notable cases, floodwaters introduced contamination into medical facilities and also caused biohazards and chemical toxins inside buildings to contaminate them further through spillage and breakage. Additional issues relate to expanding habitats of pathogens and the biting insects that spread them, or air quality impacts due to smoke, dust, mold, ozone, smog, and other factors brought on by unsustainable urban and agricultural practices, often exacerbated by climate trends. Perhaps the most obvious concerns to address are impacts related to flood-prone rivers and coastal corridors, which are projected to house the majority of global populations in coming decades and which represent a knowable footprint of low-lying lands subject to increasing storm damage. Such scenarios, let alone their tendency to increase over time due to extensive environmental changes, are not yet part of systematic management or design thinking (e.g., Arkema et al., 2013).

Appropriately, discussions have turned to mitigating climate change and to fostering resilience in the face of multiple natural hazards. Putting such deliberations into practice, however, remains far outside usual standards. Most of the allied professions involved in planning, design, regulation, and construction have a long habit of neglecting consideration of community resilience concerning direct risks to human health and safety. For example, it has long been standard to reduce flood risk for a building or road in ways that increase off-site risks, protecting against professional liability at the expense of community interest. Still less attention is granted toward the integrity of the natural environment, which provides for clean air and water supply, along with many other ecosystem services such as flood buffering in coastal and river corridors. Issues including uncertainties and potential cumulative and synergistic threats gain public policy prominence, especially in light of climate change trends. These problems, however, are not well addressed by most parties in positions to affect outcomes. The decision process is fragmented due to siloed disciplines and a focus on "first cost" rather than life cycle value, including avoided losses and general consideration of forward-looking data (e.g., sea level, temperature, rainfall intensity) and scenarios (e.g., second- and third-order effects of critical infrastructure failures) (e.g., Aerts et al., 2014). The social system in which relevant professionals practice may stifle certain aspects of progress, and overcoming these forces will require improved understanding of what works, what doesn't, why change is at times effective, and how to best replicate those successes with environmental design practices for resilient communities.

Designers increasingly prioritize sustainable development and solutions to manage society's adaptation and resilience in the face of climate change. Few professionals, however, have received training or otherwise learned satisfactory methods for meeting the technical requirements of achieving such goals. Despite advances in sustainability-oriented policy and research, change in design practice or management decisions remains sluggish in large part due to legal constraints and customs influencing designers. Notable exceptions include the promulgation of the 1994 US Green Building Council Leadership in Energy and Environmental Design (LEED) standards, and their evolving and expanding scope. However, they still represent a minuscule proportion of total project volume. The engineering community in the United States and the world has begun to discuss and endorse guidelines related to sustainability. Notably, the World Federation of Engineering Organizations, which has in recent years adopted into its charter mandates for sustainable engineering practices, strongly influenced the formulation, including metrics, of the United Nations Sustainable Development Goals (effective January 1, 2016, replacing the Millennium Development Goals). The Institute for Sustainable Infrastructure exists to guide asset owners and their design teams in pursuit of sustainable strategies, and has been training and certifying professionals and projects for 5 years. The highly trained design professionals and asset managers who would appear to be most informed, and to have the most tangible interest in guiding change, however, have not been quick to change, and often express their objection to various forms of innovation. Many issues at work within the social system of client, designer, constructor, and legal conventions dampen acceptance of innovation, favoring established and familiar methods readily fitting into the standard of care doctrine. Over the past century or more, the vast majority of development and infrastructure projects have been the culprit for unsustainable resource management and factors contributing to climate change. Much can be learned from examples where innovation has been embraced, and importantly, the very objections and points of resistance can themselves suggest specific targets for improved solutions.

An illustrative summary of the state of the design field with regard to integrating natural systems can be made through comparing and contrasting two books. *Design with Nature* (1969) was penned at the very birth of the sustainable design movement by one of its foremost leaders and teachers, Ian McHarg. A freshly written analog is the book *Building with Nature* (DeVriend and Van Koningsveld, 2012) recently published by Ecoshape, "a consortium of private partners, government agencies, and knowledge institutes operating at the nexus between nature, engineering, and society." Both books cover riparian and coastal settings; considering spatial and temporal contexts, factoring geologic processes, and biota as part of the system. However, it is how these aspects are considered and who is involved in the decision-making that differentiates the books and illuminates the authors' perspectives. Both books were written with designers in mind, but the latter embraces the importance of engagement with community members and diverse professionals.

*Design with Nature* features many analytical maps and drawings depicting the sort of spatial analysis that have since become commonplace by users of Geographic Information Systems (GIS). Indeed, many people credit McHarg for developing and popularizing the manual shape-based overlay techniques that gave rise to powerful computerized GIS analytical tools to support design decisions (Boeser, 2005). The set of environmental factors McHarg espoused reads like the chapter and verse of the National Environmental Policy Act (NEPA). Coming from education and professional practice as a landscape architect, and teaching at the University of Pennsylvania, he was not a scientist by training. However, he influenced many designers to be aware of the science underlying a project's surroundings, especially regarding dynamic settings such as barrier islands.

McHarg considered the natural process as a factor in design, whereas Ecoshape harnesses the natural process through designed elements. In its words, Ecoshape is "moving away from building in nature towards building with nature." Ecoshape also defines design guidelines based on processes that engage varied disciplines and stakeholders. Indeed, *Building with Nature*'s subtitle is "Thinking, Acting, and Interacting Differently." All the members of Ecoshape are based in the Netherlands and have gained lessons learned from recent (and past) efforts to address extreme and very real coastal risks including climate-driven threats. Ecoshape focuses on problems that include massive loss of life and compromised critical infrastructure from floods, notably during the 1950s before major sea defenses. More recently, degradation of aquatic habitats and rising costs to operate and potentially expand defenses tilted policy toward a different strategy. Newly favored solutions feature the beneficial use of dredged material for geomorphically changeable coastlines, living shorelines with native vegetation, construction of breakwaters including living oyster reefs, and others that balance public recreation, flood risk reduction, habitat improvement, and so on. Whereas McHarg articulated an ethic of stewardship, Ecoshape manifests an urgent need to address pressing problems affecting not only the environment but the viability of a nation. Specifically, Ecoshape alludes to the origin of new design ideas that can be formed based upon mimicking or cooperating with natural processes. The key benefit is that natural processes such as sediment deposition and living reefs and marshes can offer self-repair, adaptation to changing conditions, and other self-organizing attributes helpful to reducing risk, cost, environmental impacts, and other concerns.

McHarg exhorted project designers to consider issues related to community interests and the public good while Ecoshape advises engaging key stakeholders within a program of decision-making over the entire project life-cycle. Even thought-leaders such as McHarg were not adequately equipped, or perhaps inclined, to include detailed science or actual community members, or to contemplate involvement (simply consideration) beyond the design phase. In contrast, Ecoshape recognizes the importance of involving engineers, ecologists, social scientists, business, government, and citizens to engage in the project from the earliest planning through phases that may include adaptive management and abandoning unsuitable infrastructure elements. The details that differentiate the latter mode of natural-systems-based design can inform and equip practitioners with decision-support for improved sustainable development and climate resilient outcomes.

Design builds upon past work of others, performing careful analysis, and addressing newly discovered or inconsistent findings. A body of literature exists concerning interdisciplinary design research that provides theoretical framing on how research can be applied into, for, and through design, and Anderson and Shattuck (2012) offer a clear and intuitive interdisciplinary design research process:

1. State the problem
2. Analyze the context
3. Examine relevant literature, theory, and practice
4. Design a solution
5. Implement solution in authentic context
6. Iteratively evaluate, modify, and reimplement the solution

Applying this framework to contemporary design practice can provide some broad-brush insights toward identifying and implementing improvements. The problem is the customary use of maladaptive design solutions. The context involves lack of awareness to find better solutions, combined with design professionals and project owners constrained by budget, schedule, habit, and concern over potential liabilities. The relevant literature and theory span public health, asset management, ecosystem services, sustainability, resilience, design process, interdisciplinarity, and more relating to evolving practice. The solution involves putting into action (in a way that overcomes the current context of obstacles) the human–ecosystem interface spoken of often in the past where natural systems and the built environment coexist harmoniously (e.g., Odum, 1996; McHarg, 1969). To work out what strategies should be followed to achieve this demands testing and refining ideas and methods in the real world where local landscape features and functions, community preferences, engineering certification, regulatory approval, budgets, schedules, and more. Only through recursively learning and improving the solution can progress be made, and no complete validation or panaceas are likely to exist.

Decision support methodologies for integrating natural systems (or engineered systems with similar patterns of complexity and resilience) within the built environment must fit within the structure of applied project design assignments. Incorporating ecological cycles and engineering processes that resemble them such as self-sufficient photovoltaic power supply, into the built environment improve its resilience (Bridges et al., 2015). In turn, resilience can be defined as the capacity to prepare, absorb, recover, and adapt in the face of disturbances such as natural disasters, including those driven by changing climate and related conditions. Current and evolving study of resilience in engineering terms points to identifying the proper balance between redundancy, self-sufficiency, robustness, and flexibility of systems (Park et al., 2013). These characteristics at times appear counterintuitive or self-contradicting, but on closer scrutiny suggest the merit of solving design problems through multicriteria optimization, rather than the standard practice: single discipline consideration. Living systems have the added benefit of evolutionary tendencies for self-organization and self-repair that are typically absent (or costly) in their inert material counterparts. For example, a vegetated coastal bank can regrow, repair itself, and adapt to changing conditions after a major storm, but its counterpart using stone armor becomes nonfunctional after damage by impact from heavy waves or ocean debris (Barrett, 1999).

To improve design practice, we must understand patterns of misunderstanding, conflict, and unsatisfactory trade-offs that commonly contribute to maladaptive design. We must then develop, apply, and evaluate useful information, procedures, and tools that are accessible to design teams pursuing improved outcomes for sustainable development and climate resilience. This foundation, in theory, provides deeper understanding than could be obtained through evaluation of applied practice alone; experience gained in the field, including stated preferences of key decision-makers, may provide a crucible of realism (Goldsmith, 2016).

It would not be satisfying to investigate decision-support methods that may be ideal from a theoretical standpoint while failing to fit within the exigencies of professional practice. Nor would it be adequate for approaches to show promise by supporting incremental improvements while neglecting global needs. The hope is that designers can incorporate ecological cycles into the built environment:

- As a custom during routine assignments across a wide range of projects
- Linking, without harm, to existing natural systems that remain intact
- Boosting or restoring the productivity of natural systems
- Creating ecological cycles where they did not exist before, as an offset
- On a scale that matters for climate balance and community resilience

Training and standards give focused and restricted responsibilities to engineers, architects, planners, and managers who have shaped the built environment. While we now recognize the built environment to be both threatened (by increasing climate-related vulnerability) and threatening (by contributing to greenhouse gas emissions, heat island effects, and watershed degradation), there has not been a systematic retooling of the forces that shaped it. Given the known and emergent links between climate change and public health, and the disaster impacts (from floods, heatwaves, forest fires, etc.) that disproportionately affect vulnerable populations (poor, sick, aging, etc.), environmental design improvement is a policy priority, not merely a concern of property owners. How can the very same professionals whose decisions and actions over many decades have created these impacts and problems become equipped to do better? If not them, who? And if not now, when? There are large socioeconomic, ethical, ecological, and public health problems as well as potential solutions at stake.

The large scale of some problems makes it hard to conceptualize them, let alone define a ready set of remedies to solve them. Talk of vast river and coastal corridor interventions, which may require decades of multiparty cooperation, are now underway. These discussions are led by government, academic, and nongovernmental organization (NGO) thought leaders, with noteworthy examples being the design competitions launched since 2013: Rebuild by Design focusing on NY/NJ coastal resources and Changing Course focusing on the lower Mississippi River/Delta system. These efforts seek to address holistically the scale and cross-disciplinary requirements understood to be necessary to achieve long-term regional benefits using creative solutions departing from conventions and precedents. The merit of this approach, however, has not been instantly perceived by many of the people relevant to its success, in large part because of its size, complexity, and unfamiliar image. It can be useful to explore simpler examples that illustrate similar points that are more familiar to the general public, but also address public health and safety concerns, along with attempts to better incorporate sustainability and resilience.

One general form of a design standard is the prescriptive requirement; for example, government requirements concerning the positioning and power supply for smoke detectors to warn building occupants of fire hazards. In this case, the standard is obviously a good idea for improved safety, and many schools and families start young to teach their children of the need to test equipment, practice fire drills, and verify safety preparedness. But what about the similar standards for lighted exit signs and emergency lighting to guide people toward safety? Recent developments in materials, combined with experience in dangerous fires where standard systems failed to produce safe outcomes, suggest the advantage of adopting better solutions. Many professionals interested in safety and sustainability, as well as economical construction and operations, have touted the use of glow-in-the-dark signs and striping instead of hard-wired equipment that depends on electricity during the disaster event, often a faulty assumption. In normally lit buildings, this technology operates by photochemically storing low-level light for periods of over one month, then radiating visible light during darkness. Requiring no electrician to install it, and functioning during power outages (without battery backup), the material is less prone to failure and costs less. However, it does not readily fit within the defined standards of building fire codes, and it has been routinely rejected even when designers vigorously advocate for acceptance. While this example shares no outward commonality with natural systems design (other than harnessing locally available light), it does share links to sustainability and disaster resilience, and more important, it illustrates how even a seemingly simple product substitution can pose compliance and acceptance problems for designers and regulatory reviewers. The "self-organizing" aspect of emergency signs that tend to their own lighting source and need no help from electricians or the power grid is similar to vegetation, yet its virtue is "unseen" in the prescriptive design standard.

With large, intricate systems incorporating natural processes, the complexities multiply and pose further challenges to understand, evaluate, and implement. It is useful to equip designers with the means to justify design selections, and for other decision-makers to understand issues at stake. Systems-level solutions (based on natural processes or otherwise) are at a disadvantage regarding prescriptive design requirements, which inevitably cannot account for complexity or local preferences. It is useful to focus on how to anticipate and overcome hindrances that face the advantageous solutions with virtues that remain not only unseen by many decision-makers but unvalued. To accomplish this requires a cultural shift from poor interdisciplinary approaches that favor prescriptive methods to strong interdisciplinary approaches that harness, and also address, local resources, issues, concerns, and preferences.

Shortfalls related to maladaptive design solutions typically come down to important information falling between the cracks during decision-making. Many highly trained people may be involved in a project, much relevant information may be considered, and motives to attain high goals may be quite sincere, but despite the team being highly knowledgeable, at a critical handoff, continuity may be dropped somehow. Several forms of discontinuity occur, allowing failed communication, lapsed accountability, and neglected action. For instance, it is common for a decision to be considered mission critical during the design phase, yet nobody involved during construction is made aware of the issue. This pattern does not arise by accident, but through intention, convention, and age-old tradition. It may at first appear that improved coordination would be the remedy, but the gaps arise precisely due to the degree with which people embrace, preserve, defend, and adhere to their tightly defined enclosures of responsibility. Such separation is created by our most esteemed cultural institutions, notably present in the form of educational departments and distinct degrees such as architecture, engineering, business, hard science, social science, medicine, and law. People educated like this go on to obtain job roles within agencies and business entities that are similarly segmented. The granting of professional licenses and the legal administration of "standard of care" practice strongly guide the continuation of single-discipline-focused practice.

These fractured practices dominate public and private sector players in the field of project planning, finance, design, and regulatory oversight. The comfort zone of the individual is dictated by their training and work experience. Venturing beyond these norms brings discomfort and, at times, a high risk of legal ramifications. Individuals responsible for financial resources or regulatory authority seldom will heed others whose very comments or queries may appear threatening. Those responsible for keeping construction costs within budget care little for saving money within the operating cost budget; they are separate funding channels and often compete for resources and authority. The regulatory department concerned with water quality is often at odds with the one involved in air quality, and these can be sources of conflicting priorities that undermine each other's mandates, rather than balance them. A professional in structural engineering, for example, can neither afford to know nor risk not knowing all the implications of their design with regard to water resource engineering or geomorphology. A simple and safe structure that poses a low risk of professional liability may be chosen, even when it brings about unwanted impacts, so long as those impacts are not acute public safety concerns. For example, it has long been common for bridges to use fill to support approach roads to an elevated deck spanning a waterway, which encroaches on floodplains but reduces cost and complexity of bridge construction. Many bridge failures during major coastal and river storm events have involved collapses due to factors other than the integrity of the structural materials. For example, the land eroded out from underneath the approaches, scored below the support piers, or changed wholesale to bypass the bridge entirely. The result is an immediate hazard for drivers, and a long-term problem for emergency responders (fire, rescue, ambulance, etc.) and general access. Inside their disciplinary boundary, people comfortably practice what they know best, and they experience few challenges or surprises. Yet the larger issues, the spaces and cracks between and beyond the boundaries, remain neglected, with public health and environmental impacts, and with economic costs as a result. Long defined as externalities in the field of economics, similar subjects, which are routinely unaccounted for, exist in the sector of planning and design for infrastructure and development. Fortunately, progress appears within reach through understanding the causes, and better applying solutions during environmental design processes.

Each of the disciplines involved holds a deep allegiance to its own sphere of influence, but there are many levels of consequence when professionals do not think beyond conventional confines (Ulanowicz et al., 2009). For many reasons they prefer to work within the realm of that which is known, while being aware of the "known unknowns," which often are framed as the responsibility of another party. What military risk and decision scientists would refer to as "unknown unknowns" represent a body of factors that the usual mode of decision-making does not take into consideration within a professional design practice. Unknown unknowns can be critical to a project owner, or a larger community, but because they are unknown, they remain unregulated, and contractual responsibility or legal liability is not established. As a result, there are disincentives to identify or document these issues. For example, it would be difficult or impossible to address given standard budgets, schedules, and technologies, or simply due to being beyond the professionals' mental margins of training and experience; identifying questions or concerns without resolving them could establish punitive liability. By definition, resilience affords the capacity to address "surprises" that had not been "known" or systematically planned for (Park et al., 2013). Designers who are conditioned to avoid consideration of "unknown unknowns," however, remain far from adept at engaging in effective decision processes related to resilience within project solutions. Many forms of control mechanisms are in place (budget cycle controls, quality assurance reviews, zoning approval processes, environmental compliance permits, construction inspection, and more). However, none address the overall recognition, integration, and resolution of overarching factors related to resource conservation and scale and significance to natural system integrity and community resilience. Some of the largest impacts to watershed health and changing climate arise from the buildings and infrastructure created by design professionals, yet most lack training or other formal insight into the topics, rendering them "unknown unknowns" in their routine practice. In fact, only in recent decades have design professionals begun to think about these problems meaningfully in the first place, for example, linking building energy efficiency and fossil fuel consumption with climate change mitigation and resilience.

Every city and state department of transportation has established detailed standard specifications for even the most mundane-seeming items (e.g., regarding the density, material, size distribution, and allowable foreign material for rock placed as armor on banks, slopes, and channels). Any designer who applies that standard specification operates on safe ground from a liability standpoint, with construction contractors knowing precisely what material will (or will not) be acceptable, including correct pricing. Applying the living system based alternatives has been reduced to standard specifications by only a small set of agencies, leaving the onus on designers to make each and every selection and its documentation. At many points during the review process (senior engineer quality oversight, agency constructability review, contractor alternative proposals, etc.), pressure will arise to adopt normative designs, which are typically maladaptive in terms of sustainability and resilience, instead.

Many drivers and processes affect how sustainability and resilience-oriented outcomes may be integrated with design decisions of interest to business, government, and the professional disciplines that shape policies and projects. Specifically, sustainable and resilient design can be improved and realistically attained by recognizing and addressing, and in many cases harnessing, natural processes within the built environment as espoused by Odum (1996) for their "self-design" merits. Though not a panacea for addressing "unknown unknowns," "surprises," or deficits in design team role or scope, the ability of natural systems to tend to their own functions, at least in certain important ways, is attractive in the context of resilient environmental engineering solutions. At a minimum, the protection or enhancement of existing functional systems merits wider attention, especially in geomorphically dynamic and flood-prone areas. The potential to actively engage in design to restore, expand, and employ hybrid "green plus gray" solutions to engineering and construction measures for coastal and river management is moving from theory to research, and into practice.

To accomplish improved design methods on a wide scale, the evaluation and benchmarking of sustainable design practices, including those that fall short concerning natural systems within project development, become important. Perhaps more imperatively, it is necessary to note the factors that hinder performance within design teams and affect the support or interference from parties external to the design process (but whose influence matters to project outcomes). Ruckelshaus and others (2013) provide a comprehensive review of real-world decision-making related to ecosystem services, with a notable finding that though scientists may think decision tools oversimplify research, decision-makers seek even simpler decision support tools. They differentiate four pathways to produce improved outcomes for ecological services:

1. Conduct research—Produce results; publish; disseminate findings
2. Change perspectives—People become aware, understanding and discussing topic; stakeholders articulate differing ecosystem services positions; differences are revealed then mediated
3. Generate action—Present alternatives based on ecosystem services framework; considerate plans and policies; new finance mechanisms are established
4. Produce outcomes—Enhanced ecosystem services provision; improved ecological and human well-being occur

Per the Ruckelshaus framing, it is practical to focus on the third pathway related to generating action, by considering where professional design teams have accomplished project development that features natural system functions and values in order to derive improved methods for further action. It is also necessary to examine the second pathway, noting that changing the perspectives of design team members and stakeholders is needed to allay conflicts and to achieve cooperation and that appealing outcomes generate future support, even from small projects. Less directly, the first and fourth pathways also relate to this topic as research continuously contributes to the development of alternatives, and formal and informal study of project outcomes recursively informs future efforts.

The interest lies in examining what the obstacles were, how they were overcome, and whether or how generalized approaches could be formulated for application in other cases. Designers and stakeholders care about the real-world context that governs practical applications of design, construction, environmental regulations, costs, and other relevant conventions and constraints. Twenty-five years of the author's professional practice has produced the opportunity for a cycle of learning from project results as well as evolving frameworks of policy and preferences that shaped decision-making for projects. Several insights are offered to help clarify the current situation in which professionals practice:

- Community resilience is greatly affected, for better and for worse, by practices of design professionals including planners, engineers, architects, landscape architects, urban designers, and the allied professionals in natural sciences, social sciences, economics, finance, insurance, law, and public health.
- Designers can, but generally do not, provide leadership to improvise exchange of information with researchers in order to drive innovative approaches.
- The defined boundaries of professional responsibility are real, and still generate conflicts, though recent schooling has produced willingness to connect across boundaries.
- Science is critical for design practitioners and stakeholders, but the challenge is simplification in ways that support the design process and associated internal and external communication.
- Life cycle project costs, impacts, and vulnerabilities are important, yet designers are trained that their role ends when plans and specifications are complete, and few understand operational details of structures, let alone nature.
- Design procedures create routines, in turn providing efficiency and clarity, as well as resistance to change; adapting existing procedures creates a faster buy-in.
- The interaction process between team members and stakeholders is seldom part of design training, and therefore practice. Its absence or weakness often damages outcomes, with public health impacts often remaining outside the set of factors considered.

Moncaster and others (2010) identified the critical gap between academic research and sustainable engineering practitioners, and advocated for improved knowledge exchange. Knowledge exchange must be demonstrably practical for all parties, commencing with insights gained from professional practice, and reflecting research investigations undertaken iteratively over decades, all within a context of evolving theory and increasingly comprehensive literature and performance data. Practical items may appear lacking in rigor to academic researchers, but hold great interest to policy-makers, practicing professionals, and the wider community concerned with project impacts and benefits. Lessons learned, patterns observed, constraints and resistance encountered, and gaps identified are especially helpful. Procedures that highlight the merits of working across disciplinary boundaries can help initiate effective team-based design. Procedures that support the efficient development, analysis, and application of site-specific and context-sensitive project information can improve the willingness and the capacity to perform work using interdisciplinary design. Effective design teams can become better equipped to develop the business case for greater environmental and social responsibility.

It is not adequate, however, to focus only on the design process, as the characteristics of the design solutions matter long after the designers depart the scene. To support sustainable development and climate resilience objectives, functional characteristics of the built solutions will ideally support efficiency, self-sufficiency, self-repair, adaptation, and resilience. In short, self-organizing natural systems are the main tool in the process, and designer methods that align to produce these systems set the stage for the ongoing performance of these systems. This topic is highly relevant to design professionals, but also has value for allied professionals and other stakeholders. The same measures, when viewed within a regional or global context, may be integrated (with positive versus conventionally negative impacts) as solutions for watershed health and global climate functions, transforming projects aimed at water quality compliance or hurricane damage reduction into watershed restoration and carbon solutions. The functions of the natural systems may become better addressed through thoughtful integration within (and outside of) design teams, imparting knowledge from multiple disciplines to frame, analyze, and solve design problems.

It is not compelling to merely examine whether designers can do this or how communities may benefit. What ultimately holds interest is to examine the factors, processes, and decision support tools that might better allow designers to do it in the following ways:

- As a custom during routine assignments across a wide range of projects
- Linking, without harm, to existing natural systems that remain intact
- Boosting or restoring the productivity of natural systems
- Creating ecological cycles where they did not exist before, as an offset
- On a scale that matters for climate balance and community resilience

Given the innate complexity of this inquiry, limitations exist governing the structure of comparison between projects with differing contexts. From a portfolio of over $15 billion in project implementation costs, two projects are selected that illustrate the application of the interdisciplinary process for the design of environmental engineering solutions at the scales of a large urban neighborhood and a multicommunity region. The insights are not based on mere theory, or trivial or marginal professional practice. The challenges related to designer and stakeholder roles were not only real, they received stringent attention. These projects form a high-value portfolio from which to extract relevant illustrative observations and recommendations.

The city of Cambridge, Massachusetts, developed a creative plan located within Olmsted's Emerald Necklace park and greenway system to mitigate combined sewer overflows. During large rainfall events, raw sewage was being discharged into waterways flowing through public parks, causing exposure to many pathogens as well as nutrient overload. Considered the crown jewel of the legally mandated Boston Harbor cleanup project, one of the United State's most complex and challenging storm water treatment wetlands, was selected as a cost-effective climate-resilient solution. The designers navigated complex regulatory approval for the integrated approach to modifying traditional piping and adding large created wetlands plus other habitat features and public amenities. By rerouting runoff into a carefully sited and engineered basin in the Alewife Reservation, storm water was effectively separated, stored, treated, and released with improved peak and base flow patterns that address chronic local flooding as well as court-mandated water quality requirements. Public fears that wetlands could breed mosquitos were dispelled through careful design (to foster mosquito-eating aquatic life) and monitoring (to document that the system bred very few mosquitos, unlike standard local catch basins). The multistage treatment train contained within sinuous vegetated earthen berms removes sediment; provides biological filtration to degrade harmful chemicals; and promotes breakdown and removal of oil, grease, nitrogen, pathogens, and more. Differing from prior approaches that stored water in large underground tanks for later pumping and treating at Boston's Deer Island facility at great energy cost, this natural approach is expected to store carbon in soils and biomass, and to educate park-goers on community resilience topics for generations to come. It is designed to function passively at all times, so during power outages public works crews are free to address other community needs. Maintenance is similar to standard park land and storm water pipe/basin procedures and expenses, but with added community benefits.

The Greater New Orleans (Louisiana) Hurricane and Storm Damage Risk Reduction System (HSDRRS) formulated a regional scale infrastructure system after Katrina that factored in trends and uncertainties surrounding future climate change and extreme land subsidence. Only a regional approach could ensure reasonably reliable continuity of operations for critical infrastructure, distinguishing the project from past efforts. A network of existing, upgraded, and new floodwalls, gates, pumps, and ancillary structures provided a truly integrated system intended to function adaptably facing multiple scenarios. Failings noted during Katrina were addressed through measures including redundant backup power supply, safe houses for workers to shelter in-place at system controls, and key elements elevated to withstand flooding while providing continued operation to allow flood-water removal. Recognizing hard infrastructure could perform much better long term while surrounded and shielded by healthy coastal wetlands, and protective landforms such as barrier islands were included as core functional elements of the design. The engineering of the structures also accounted for natural-systems-based solutions for synergistic performance based on specialized knowledge of the physical properties of vegetation, including in combination with construction materials. Wetlands, beaches, and dunes were constructed using dredged sediments, often with reinforcement by "hard" materials. Levees, floodwalls, and the world's largest surge barrier serve to keep damaging saltwater away from sensitive vegetation, in addition to its main function to prevent downtown flooding. The seamless integration of green and hard elements blurs the boundary between natural and built, and harnesses natural processes to support and maintain the coastal region, and to best adapt to future changes and impacts. Together, natural landforms, healthy ecosystems, and built structures are capable of mitigating storm damage based upon multiple lines of defense, delivering sustainable infrastructure solutions many cities facing the need for climate change adaptation can emulate. Surprisingly, the project went from conceptual design shortly after Katrina's 2005 impact to effective performance prior to the 2012 impact of Hurricane Isaac. Despite common assumptions that large projects always require long timelines for completion, the work was executed rapidly enough to prevent a repeat of similar damages and is understood to have avoided losses estimated at over $30 billion.

Integrating natural systems with design of the built environment can offer important benefits to community resilience at many levels. The types and levels of integration necessary and beneficial within an engineering context are manifold. At the most literal level, physically incorporating natural materials, native plants, and ecological processes within engineered construction measures achieves pragmatic integration. While this alone is not standard or common, it is gaining popularity from a policy standpoint.

At the level of integrating professional disciplines, challenges exist preventing straightforward collaboration and agreement. In the common pursuit of applying vegetation within engineering measures that demand multiple areas of knowledge, conflicts frequently arise during design practice between professionals. For instance, soil scientists have strong motives to limit compaction of soils to retain agronomic properties and support root growth and development, whereas geotechnical engineers have strong motives to maximize soil compaction to obtain the most consistent and physically durable properties, despite being incompatible with root growth and other ecological functions. In most cases, these professionals lack awareness of the other parties' motives or standard practices, yet the engineers normally trump the decision process, being the ones called upon to certify and seal the design. To span these two disciplines, methods must reconcile apparent conflicts while addressing professional standards of practice. For those projects drawing on still more disciplines, the challenges multiply, but still may be resolved through transparent and accountable decision processes supported by appropriate tools.

At the level of integrating natural-systems-based measures within larger landscape systems, harmonizing approaches that address sustainability and climate goals comprehensively requires efficiently using available resources while minimizing regressive impacts. Most approaches to engineering solutions to river erosion or climate vulnerability rely on materials and methods that generate climate impacts. For example, in the name of urban flood control, ecologically productive riparian corridor forest communities that efficiently capture and store carbon from the atmosphere are routinely displaced by concrete structures. In doing so, the production of cement releases greenhouse gas emissions both directly and indirectly, and the river corridors no longer function to aid greenhouse gas management. In many cases, the flood problem is also displaced to a nonurban zone, rather than being systematically solved. By considering a decision process that accounts for photosynthetic potential and carbon storage, decision support methods could better aid designers. However, it remains uncommon for design professionals (and the clients who hire them) to define specific issues, targets, and methods for addressing these issues. Examples where projects have successfully implemented comprehensive solutions share the hallmark of a strong interdisciplinary design process.

At the level of integrating the multiple parties, both internal and external to the design team, a common challenge is resolving conflicts to enable effective and creative collaboration. Real and perceived conflicts cause time, money, and goodwill to be spent fruitlessly. Identifying shared interests and understanding the context within environmental, social, and economic systems of many scales allows groups to solve problems while minimizing externalities that commonly incite conflict. Applying a natural systems approach allows adequate consideration of solutions that minimize externalities and avoid/resolve conflicts. Equipped with established methods, design teams could expend fewer resources and achieve more objectives with lower impacts, thereby allowing a greater return on investment, consistent with community interests. Achieving this requires suitable engagement of stakeholders to broadly identify the range of interests, values, and preferences to consider, and to structure decisions to address these, while avoiding maladaptive approaches and undesired impacts.

Leadership toward facilitating such a process may be initiated by virtually any party, from formally charged engineering professionals, to special interest activists, to community representatives, to allied professionals informed and concerned about subject matter underlying public health and community resilience. Engagement between project owners/investors (public agencies, private corporations, or civic institutions) and parties capable of (or vested in) sharing inputs about key concerns and desired outcomes during early planning can support having important issues identified, prioritized, and considered in the first place. Best practices throughout design and further project implementation can ensure continuous attention to important issues that are otherwise routinely disregarded in favor of habits and customs. The roadmap for resilient environmental design begins with community input and ends with ongoing ecological processes aligned with withstanding and recovering from disaster impacts.

# 10

## The Epoch of Smart Cities and Innovation

Nathalie Crutzen with Cesar Marolla

## Smart Cities: Context and Definitions

Our planet is indeed facing significant demographic challenges. Earth's population has increased exponentially, and it continues to rise. Today, more than half of the population lives in (mega)cities (Nam and Pardo, 2011). This percentage jumps to more than 66% in the European Union (Eurostat, 2014). Some analysts predict that 70% of the world population will be living in cities by 2050 (*The UN Global Compact-Accenture CEO Study On Sustainability 2013*, 2013). The fast growth of the urban population implies numerous economic and societal challenges in domains like mobility, housing, employment, education, culture, security, and natural resource management: water, waste and energy (Nam and Pardo, 2011).

Climate change can be perceived as a cause, consequence, or both for this increasing urban population. On one hand, some impacts of climate change such as unprecedented floods, desertification, and storms, might be the catalyst for the migration of people towards cities where life is viewed as more comfortable. On the other hand, this increasing urban population (in megacities in particular) contributes to the pollution of our planet (water, soil, air) and drains natural resources. This pollution has adverse consequences for the megacities' dwellers and exacerbates climate change implications on public health.

In this context, it is crucial to develop in-depth thinking, strategies, and actions for sustainable development of our urban ecosystems and a better quality of life for its citizens. It is essential that the leaders of the city develop and implement long-term sustainable strategies to mitigate their impact on climate change and to create adequate economic and societal environments, where citizens, companies, and public authorities can live, work, and interact. Alongside a top-down approach, these leaders should also provide adequate conditions to support the development of bottom-up initiatives that will contribute to the dynamics in a sustainable manner.

The concept of a "smart city" emerges more and more; it limits the impacts of urban population growth and climate change by finding innovative solutions to meet these challenges.

Interest in smart cities is tremendous and increasing at the international, national, and regional levels (Chourabi et al., 2012). The multiplication of platforms and other initiatives demonstrate growing interest all over the world.

Currently, only a limited number of scientific publications are dedicated to smart cities (Chourabi et al., 2012), especially in management science. Most of the existing initiatives, publications, and definitions were focused on very technical solutions in particular domains (mobility, energy, water, etc.) without proposing a real long-term strategic vision and business or managerial thinking on these questions (business models, financing, stakeholders' dynamic, etc.).

The concept of the smart city is still emerging and needs more elucidation (Nam and Pardo, 2011). The smart city has no generally accepted definition in either literature or practice. This situation is confused because many other terms or labels are used (e.g., intelligent cities, digital cities, wired cities, sustainable cities), sometimes synonymously but not always (Hollands, 2008).

Some examples of different published definitions of the concept follow:

- "Territories with high capacity for learning and innovation, which is built-in the creativity of their population, their institutions of knowledge creation, and their digital infrastructure for communication and knowledge management" (Komninos, 2002).

- "One of the key elements which stands out in the smart (intelligent) city literature is the utilization of networked infrastructures to improve economic and political efficiency and enable social, cultural and urban development" (Hollands, 2008).

- "The use of ICT [makes] the critical infrastructure components and services of a city—which include city administration, education, healthcare, public safety, real estate, transportation, and utilities—more intelligent, interconnected, and efficient" (Washburn and Sindhu, 2009).

- "A city is smart when investments in human and social capital and traditional and modern communication infrastructure fuel sustainable economic growth and a high quality of life, with a wise management of natural resources, through participatory governance" (Caragliu, Del Bo, and Nijkamp, 2009).

- "Any adequate model for the Smart City must therefore also focus on the Smartness of its citizens and communities and on their well-being and quality of life, as well as encourage the processes that make cities important to people and which might well sustain very different—sometimes conflicting—activities" (Haque, 2012).

- "The idea of Smart Cities is the creation and connection of human capital, social capital and information and Communication technology (ICT) infrastructure in order to generate greater and more sustainable economic development and a better quality of life. A Smart City is a city seeking to address public issues through ICT-based solutions on the basis of a multi-stakeholder, municipally based partnership" (European Parliament, 2014).

These solutions are developed and refined through smart city initiatives, either as discrete projects or (more usually) as a network of overlapping activities. Smart cities have been further defined along six axes:

- Smart economy
- Smart mobility
- Smart environment
- Smart people
- Smart living
- Smart governance

In the very recent past, technology-centered approaches were predominant; over time, broader definitions and more global approaches including the three pillars of sustainability and human and social capital have emerged.

Based on previous literature, the Smart City Institute proposes a broader definition of a smart city. A smart city is a multistakeholders ecosystem (composed of local governments, citizens' associations, multinational and local businesses, universities, international institutions, etc.). These stakeholders are engaged in a sustainability strategy using technologies (ICT, engineering, hybrid technologies) as enablers for all spheres of sustainable development (Allwinkle and Cruickshank, 2011; Belissent, 2011; Caragliu, Del Bo, and Nijkamp, 2011; Chourabi et al., 2012; Hollands, 2008; Kourtit and Nijkamp, 2012; Kourtit et al., 2012; Lombardi et al., 2012; Nam and Pardo, 2011).

This approach implies the progressive development of a common strategic vision and the development of concrete initiatives in various domains (smart mobility, environment, economy, living, people, and governance) in order to generate sustainable economic development and to offer a better quality of life along with a wise management of natural resources.

In addition to this strategic perspective, smart cities also require the development and diffusion of new business models that will contribute successfully to their transition toward sustainability and innovative financing instruments, as well as a good understanding of specific stakeholders' dynamics.

## The Smart City Institute at HEC-Management School of the University of Liege

Considering the demographic and climate change challenges our planet is facing, the Smart City Institute (SCI) was launched at HEC-Management School of the University of Liege (Belgium) (HEC-ULg) in January 2015. The SCI is an academic institute dedicated to the thematics of smart cities. It aims at stimulating research, teaching, innovation, and entrepreneurship in the field of the smart city. Unlike other existing and famous academic centers that adopted predominantly technological approaches, the SCI proposes to tackle this phenomenon from a business/managerial angle while collaborating with scholars and practitioners from other disciplines (including political sciences, engineering, urbanism, and geography). The mission of the Smart City Institute is to "contribute to the general development of smart cities by training future managers, developing research, entrepreneurship and innovation as well as facilitating sustainable value creation between actors of smart ecosystems utilizing several tools such as networking and access to multidisciplinary skills and the most innovative technologies."

## Sustainable Development and Smart Cities

An ideal goal for any city is to use fewer resources, particularly in megacities where the intensive use of natural resources leads to scarcity and lack of essential services. A smart city is an extension of a sustainable city; it creates the most benefit while minimizing the impacts produced by utilization.

The goal of operating the megacity's systems efficiently is the basis for sustainable development and paves the way to a resilient megacity's environment. Reducing energy consumption and costs, and improving networks and suppliers are some of the goals of sustainable management while planning and adapting to a changing environment. Sustainable development is now enshrined on the masthead of world leaders' agendas as the path to poverty reduction and sustainable societies. Locality is the result of using information and telecommunication technologies (ICTs) in all aspects of life, and smart cities use smart systems to improve the quality of life and the efficient use of resources to become prosperous, improve the local economy, and minimize environmental degradation and impacts on sustainable development.

Megacities' explosive economic and population growth and the shift in and expansion of urban sprawl in the emerging economies will create many opportunities for companies, in particular for the ICT sector. The world's largest cities, particularly in developing countries, will become the largest markets for existing premium products and technologies, while the developed countries–megacities will witness a trend in sustainable measures (Frost and Sullivan, 2010).

Planning and a management vision that promotes sustainable urban design should guide development and create sustainability for megacities. Smart solutions help megacity development by encouraging the efficient use of nonrenewable resource solutions and the efficient use of smart solutions to produce renewable resources. Smart solutions provide improvements in the quality of life as well as efficient consumption of natural resources now that issues related to urbanization are becoming increasingly important (Batagan, 2011).

---

## Health Systems and Organizational Performance

Health care systems are already benefiting from smart solutions and ICT innovation with improved organizational performance and patient care. Some of the advantages are real-time visibility in operations, wider ranging samples to achieve more medical breakthroughs, and applications for intelligent health systems including data integration and its focus on the patient. The efficient use of resources minimizes waste, reduces energy consumption, provides better care and services, and interconnects health providers and patients. Smart solutions for a higher quality of life should guide sustainability development. Sustainable development is contingent on the open access to smart solutions and a competitive economy and an economy based on knowledge and innovation (Batagan, 2011).

# Section III

# Global Health Risks, Facts, and Challenges of Climate Change Implication

# 11

## Global Health Risks, the Urban Poor, and Climate Change Impacts

In 2009, one of the world's leading general medical journals, *The Lancet*, and University College London (UCL) Institute for Global Health produced a report on the work of a yearlong commission. It opens with the following statement: "Climate Change is the biggest global health threat of the 21st century" (Costello et al., 2009). Climate has always affected our health, and the growing acknowledgment of climate change and its consequences creates more interest in its impacts. Various factors influence the results: population growth, urbanization, land use reduction, and freshwater resources. These elements affect the community's health and magnify the impacts of climate change events (Haines, 2006).

The urgency of this threat shows the importance of focusing on the impact that a changing climate has on health. First and most important, the poorest people in the world will suffer the worst consequences. The rural population's migration to major cities creates a steady growth in urban settlements with high density, human-created structures that exacerbate the health-related problems of climate change. As the climate changes, health issues faced by the urban poor could result in entire communities being devastated, because the destitute tend to live in informal settlements that are more defenseless against drastic weather events. Climate change will increase the vulnerability of the urban poor due to reduced food security; compromised drinking water; and water-rodent-borne diseases associated with floods, droughts, and the correlation of high temperatures and heat stress. Impacts on the urban poor will widen the social and economic disparity among the population and will create a gradual problem that will affect future generations' health and increase their risk factors (Costello et al., 2009).

## Climate Change Inequalities

Climate change impacts on human health are acutely inequitable. Climate change is a threat to global public health, and the poorest populations are exposed to greater risks. Poor populations living in major cities with rapid economic development will be most susceptible to climate change effects (Campbell-Lendrum et al., 2007). The susceptibility factors of the urban poor, particularly in developing countries, are location, the effects of the urban heat island, air pollution, growing population, and lack of proper sanitation (Campbell-Lendrum et al., 2007).

## Climate Change and Public Health Impacts

According to the World Health Organization (WHO), climate change is estimated to have caused over 150,000 deaths annually since the 1970s. Higher temperatures and changing rainfall patterns resulting from global climate change also may influence transmission patterns for many diseases, including water-related diseases, such as diarrhea, and vector-borne infections, including malaria. Furthermore, climate change could have far-reaching effects on patterns of food production, which can have health impacts in terms of rates of malnutrition (WHO, 2013a).

## Health Risk

The climate has adverse consequences for human health, intensifying the risks to the population. By definition, health risk is the state of complete physical, mental, and social well-being, and not merely an absence of disease or infirmity (WHO, 1996). This definition can help us to understand health risks and have a clearer picture of what we are trying to accomplish considering the inverse definition would be "the presence of a disease and lack of completeness in physical, mental, and social/emotional well-being tantamount to health risk" (Omoruyi and Kunle, 2011). The international frameworks that incorporate best practices for organizations—ISO 31000:2009 (risk management) and ISO 22301:2012 (business continuity management)—define risk as the "effect of uncertainty on goals." This effect, which represents a deviation from the expected, can be positive or negative. Risk is usually characterized by events and the consequences of those events. Uncertainty is a lack of information that makes it difficult to understand the event, its consequences, or its likelihood (Pojasek, 2013a). Health risk is the population's susceptibility and degree of exposure to hazards. Health risks can be described as "chance of loss and/or variability from health and worsening of ill health" (Omoruyi and Kunle, 2011).

## The World Turns Urban

Significant steps toward a solution are urgently needed to diminish the health risks of the urban poor due to climate change, according to the latest United Nations report. Between 2007 and 2050, the urban areas of the developing world are expected to absorb an additional 3.1 billion people while the overall population will grow by just 2.5 billion people. The rapid and unplanned expansion of megacities exposes a greater number of urban residents to climate change health risks and increases their vulnerability to extreme weather events. The health impacts will depend on the region, susceptibility of populations, and the capability of societies to adapt and mitigate the impacts (Dickson et al., 2011).

The challenge for megacities is to develop an effective urban risk management framework that provides guidance for maintaining and delivering essential services and lessening the effects of climate on health. Cities need to consider the issues of climate change and urban health by evaluating related risks to identify feasible measures in their planning and management processes (Dickson et al., 2011).

Susceptibility to climate change and its effects on health increase as the concentration of economic activity and population density worsens the health situation in the world's major cities. The impacts of climate change on health are magnified on urban populations (Feiden, 2011). The capacity of the urban poor to adapt is weak in comparison because of inadequate housing, poor nutrition, overcrowded living, and population displacement. Resources and information are scarce and the urban poor cannot respond efficiently in order to take actions to mitigate climate change effects; this situation creates a gradual exposure to health risks. The poor population's vulnerability will be exacerbated by exposure to severe weather effects and lack of ability to adapt to climate change (Haines, 2006).

Rising sea levels make cities vulnerable to drastic environmental changes and exacerbate the economic risks faced by poor residents. Inland megacities also have susceptibilities to climate change, including urban settlements on steep slopes in hazard-prone areas, and the heat island effects (United Nations Habitat, 2011b). Sea-level rise and other coastal implications, such as changes in storm frequency, put populations at risk of climate change effects. The rapid urbanization experienced in the 20th century and the expansion observed today has already created 20 coastal megacities. In contrast, in the 1990s there were a total of seven coastal megacities in Asia (excluding Japan) and two in South America (Nicholls, 2004). The fragility of the sedimentary strata where the megacities are located is a concern that adds to climate change impacts on urban populations. The repercussions of climate change on megacities are varied and require each city to have an independent assessment to mitigate and adapt to climate change (Nicholls, 2004).

Among the cities in the case study, the city of New York is vulnerable to midlatitude cyclones and nor'easters, which reach a peak from November to April. Coastal erosion is increasing and makes the city defenseless against coastal flooding. Cyclones and hurricanes have the potential to reach New York City and expose the urban population at a high risk of disaster (Fifth Urban Research Symposium, 2009).

Rio de Janeiro has a climate change factor that other coastal cities share. The rise in sea level will have a variety of impacts (e.g., flooding, sanitation issues, quality and quantity of water resources) (Rosenzweig et al., 2010). Los Angeles has developed a comprehensive adaptation and mitigation plan, and conducted a pioneer study of the impacts of climate change in the county; it can serve as an example to other megacities around the world. These initiatives help to create a solid strategy to address the cost of inaction to climate change impacts, and the focus is on the health effects of severe climatic events and preventive actions to reduce the burden of disease. Lastly, the research focuses on the city of Beijing, assessing heat-related mortalities, the urban heat island effect on its residents, and air pollution issues that are becoming a major concern for climate change mitigation strategists in this megacity, all affecting the health of the urban poor in particular. The local government involvement in developing a risk management plan is critical to improving the resilience of the urban poor (Rosenzweig et al., 2010). This research presents a risk management framework, the standard ISO 31000, and a business continuity plan, which provides additional insights as to how major cities can prepare, develop, and implement a risk management structure to deal with climate change issues.

## Heat Waves in Europe

Many different countries and regions around the world have experienced climate change impacts on human health. France and Italy lost more than 14,000 lives due to a heat wave across Western Europe in the summer of 2003 (Wang and Charneides, 2005). Heat waves are just one example of the harms climate change causes to human health. The range of disease-carrying insects will expand, affecting populations that have never been exposed to those factors. Crop productivity will decrease, creating a parallel increase effect in malnutrition (Heinzerling, 2007).

Paris suffered a devastating heat wave from August 1 to August 14, 2003, where the maximum and minimum temperatures reached unprecedented highs. The high temperatures were not accompanied with high relative humidity, as is usually reported, and on August 4, 300 excess deaths were reported. The reported deaths gradually increased until August 12, reaching 2,000 per day, and then quickly stopped within a few days. The death toll in the month of August reached to 14,800, which is equivalent to a 60% increase of expected mortality in France (Dhainaut et al., 2004).

## Climate Change and the Effects on the Environment and Health

The effects of climate change on the environment have an impact on public health. They include compromised water supplies, change in air quality, food security, and direct effects on vectors (Portier et al., 2010).

### Change in Water Supply

The change in rainfall will increase the severity and occurrence of droughts in some regions and floods in others. These events will impact the vector population, and water scarcity will affect crops and pasture yields. Change in precipitation patterns could change the population density of vector-borne diseases, infecting inhabitants in greater numbers and affecting the health of entire communities. Roads, schools, power supplies, and houses are affected by flooding, which destroys essential infrastructure to support communities and cities (Portier et al., 2010).

Storm water discharges are damaged by heavy rains, putting waste into the water supply. Cities near the coast will be affected by rising sea levels, which will result in salination of freshwater supplies, loss of land productivity, and a change in the breeding of coastal dwelling mosquitoes (Epstein, 2001).

In Africa alone, by 2020, between 75 million and 250 million people are projected to be exposed to an increase of water stress due to climate change. Water security problems are also projected to intensify by the year 2030 in southern and eastern Australia. In the course of the century, water supplies stored in glaciers and snow cover are projected to decline, reducing water availability in regions supplied by meltwater from major mountain ranges (such as the Himalayas in Asia and the Andes in Latin America). In North America, the decreased snowpack in western mountains is projected to cause more winter flooding and reduce summer flows, creating competition for overallocated water resources (Van Ypersele, 2007).

Freshwater availability in Central, South, East, and Southeast Asia, particularly in large river basins, is projected to decrease due to climate change, which could, in combination with other factors, adversely affect more than a billion people by the 2050s (Van Ypersele, 2007).

Figure 11.1 shows adaptation scenarios for California's urban and agricultural water use. There is a 15% reduction from conservation and efficiency

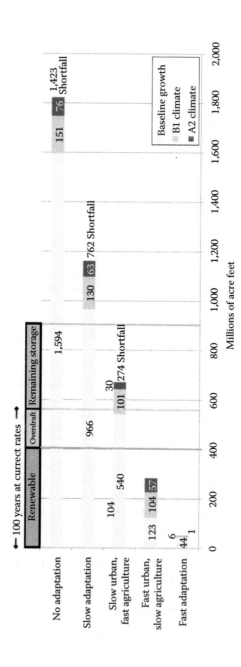

**FIGURE 11.1**
California's projected groundwater extraction plus shortfall, 2010 to 2110. (From Ackerman, F. and Stanton, E., *The last drop: Climate change and the southwest water crisis. Stockholm Environment Institute–U.S. Center, 6–7,* 2011.)

on the slow urban adaptation scenario in addition to a 20% price increase by 2030. A 5% reduction in water use and a 10% price increase is projected by a slow agricultural adaptation through 2030. According to projections, the savings from slow urban and agricultural adaptation combined are not enough to reduce California's water use to the sustainable level, even without the impacts of climate change (Ackerman and Stanton, 2011).

## Change in Air Quality

Higher temperatures change air pollutants. For example, higher temperatures affect the range and concentration of pollen, affecting a larger range of the population (Portier et al., 2010). Los Angeles remains the city with the worst ozone pollution problem but reported its fewest unhealthy ozone days since the American Lung Association State of the Air reports began in 2000. Table 11.1 shows the most ozone-polluted cities in the United States; it illustrates cities in California leading the list in 2016. The statistics put emphasis on the correlation between poverty and health outcomes (American Lung Association, 2016).

---

## Public Health Burden of PM2.5 and Ozone

In general, the percentage of all deaths due to PM2.5 exposure is higher than it is for ozone. Southern California has the highest percentage of PM2.5-related deaths among the United States population. If the air quality incrementally improves, the number of deaths related to PM2.5 and ozone decreases. For example, 23,000 PM2.5-related mortalities would be avoided as a result of lowering 2005 annual mean PM2.5 levels down to 10 µg/m³ (micrograms per cubic meter of air) nationwide. The estimate concludes that approximately 80,000 premature mortalities would be preventable by lowering PM2.5 levels to 5 µg/m³ on a national scale (Fann et al., 2011).

The U.S. Environmental Protection Agency (2012) has taken important measures to protect the health of U.S. residents, and it has strengthened the annual health National Ambient Air Quality Standard for fine particles (PM2.5) to 12.0 µg/m³. The agency also retained the existing standards for coarse particle pollution (PM10).

**TABLE 11.1**

People at Risk in 25 U.S. Cities Most Polluted by Year-Round Particle Pollution (Annual PM2.5)

| 2016 Rank[1] | Metropolitan Statistical Areas | Total Population[2] | Under 18[3] | 65 and Over[3] | Pediatric Asthma[4,6] | Adult Asthma[5,6] | COPD[7] | CV Disease[8] | Diabetes[9] | Poverty[10] |
|---|---|---|---|---|---|---|---|---|---|---|
| 1 | Bakersfield, CA | 874,589 | 257,512 | 86,198 | 22,811 | 47,274 | 27,545 | 39,611 | 58,509 | 206,604 |
| 2 | Visalia–Porterville–Hanford, CA | 608,467 | 186,159 | 61,302 | 16,490 | 32,302 | 18,893 | 27,286 | 39,992 | 160,479 |
| 3 | Fresno–Madera, CA | 1,120,522 | 321,538 | 127,627 | 28,482 | 61,434 | 37,066 | 54,190 | 78,465 | 293,929 |
| 4 | Los Angeles–Long Beach, CA | 18,550,288 | 4,419,138 | 2,287,192 | 391,452 | 1,093,121 | 670,009 | 981,745 | 1,425,473 | 3,174,300 |
| 5 | El Centro, CA | 179,091 | 51,111 | 21,523 | 4,527 | 9,863 | 6,046 | 8,897 | 12,791 | 40,162 |
| 6 | Modesto–Merced, CA | 798,350 | 225,241 | 92,260 | 19,952 | 44,214 | 26,914 | 39,399 | 57,132 | 160,041 |
| 7 | San Jose–San Francisco–Oakland, CA | 8,607,423 | 1,876,296 | 1,168,168 | 166,204 | 523,893 | 330,069 | 488,003 | 703,447 | 968,270 |
| 8 | Pittsburgh–New Castle–Weirton, PA—OH—WV | 2,653,781 | 512,313 | 489,155 | 55,262 | 210,546 | 154,349 | 218,588 | 249,655 | 331,578 |
| 9 | Harrisburg–York–Lebanon, PA | 1,239,677 | 271,569 | 204,056 | 29,398 | 95,249 | 66,506 | 94,211 | 108,812 | 129,647 |
| 10 | Louisville/Jefferson County–Elizabethtown–Madison, KY–IN | 1,498,593 | 348,103 | 213,057 | 35,700 | 134,900 | 132,472 | 132,990 | 138,376 | 213,396 |
| 11 | Reno–Carson City–Fernley, NV | 597,837 | 130,592 | 97,747 | 8,848 | 38,360 | 34,676 | 45,621 | 47,522 | 89,277 |
| 12 | Lancaster, PA | 533,320 | 128,671 | 87,385 | 13,929 | 39,794 | 27,486 | 39,175 | 44,979 | 54,499 |
| 13 | El Centro, CA | 179,091 | 51,111 | 21,523 | 4,527 | 9,863 | 6,046 | 8,897 | 12,791 | 40,162 |
| 14 | Pittsburgh–New Castle–Weirton, PA—OH—WV | 2,653,781 | 512,313 | 489,155 | 55,262 | 210,546 | 154,349 | 218,588 | 249,655 | 331,578 |
| 15 | Yakima, WA | 247,687 | 73,891 | 31,719 | 4,826 | 16,075 | 10,398 | 12,998 | 14,992 | 50,044 |
| 16 | Anchorage, AK | 398,892 | 101,730 | 36,091 | 9,374 | 23,752 | 12,587 | 16,994 | 20,760 | 39,450 |
| 17 | Sacramento–Roseville, CA | 2,513,103 | 592,935 | 358,196 | 52,523 | 149,894 | 96,523 | 144,007 | 205,390 | 397,024 |
| 18 | Philadelphia–Reading–Camden, PA—NJ–DE–MD | 7,164,790 | 1,601,349 | 1,058,447 | 164,662 | 520,226 | 350,165 | 491,940 | 577,817 | 950,284 |

*(Continued)*

**TABLE 11.1 (CONTINUED)**

People at Risk in 25 U.S. Cities Most Polluted by Year-Round Particle Pollution (Annual PM2.5)

| 2016 Rank[1] | Metropolitan Statistical Areas | Total Population[2] | Under 18[3] | 65 and Over[3] | Pediatric Asthma[4,6] | Adult Asthma[5,6] | COPD[7] | CV Disease[8] | Diabetes[9] | Poverty[10] |
|---|---|---|---|---|---|---|---|---|---|---|
| 19 | Harrisburg–York–Lebanon, PA | 1,239,677 | 271,569 | 204,056 | 29,398 | 95,249 | 66,506 | 94,211 | 108,812 | 129,647 |
| 20 | El Paso–Las Cruces, TX—NM | 1,050,374 | 290,708 | 124,863 | 20,269 | 55,486 | 39,945 | 58,111 | 81,066 | 250,142 |
| 21 | Eugene, OR | 358,337 | 68,413 | 62,334 | 4,963 | 29,455 | 16,575 | 24,260 | 26,412 | 64,722 |
| 22 | South Bend–Elkhart-Mishawaka, IN—MI | 723,537 | 178,540 | 110,538 | 15,281 | 58,635 | 48,571 | 53,112 | 58,902 | 111,135 |
| 23 | Phoenix–Mesa–Scottsdale, AZ | 4,489,109 | 1,121,933 | 638,383 | 122,364 | 325,041 | 226,682 | 264,470 | 327,660 | 753,716 |
| 24 | New York–Newark, NY—NJ—CT—PA | 23,632,722 | 5,198,379 | 3,383,979 | 473,026 | 1,812,756 | 1,039,620 | 1,392,285 | 1,785,585 | 3,281,939 |
| 25 | Medford–Grants Pass, OR | 293,886 | 60,420 | 63,154 | 4,383 | 23,312 | 14,641 | 22,447 | 24,063 | 54,487 |

*Source:* American Lung Association, 2016, *State of the Air*, p. 15.

[1] Cities are ranked using the highest design value for any county within that combined or metropolitan statistical area.
[2] Total population represents the at-risk populations for all counties within the respective combined or metropolitan statistical area.
[3] Those under 18 and 65 and over are vulnerable to PM2.5 and are, therefore, included. They should not be used as population denominators for disease estimates.
[4] Pediatric asthma estimates are for those under 18 years of age and represent the estimated number of people who had asthma in 2014 based on state rates (Behavioral Risk Factor Surveillance System [BRFSS]) applied to population estimates (U.S. Census).
[5] Adult asthma estimates are for those 18 years and older and represent the estimated number of people who had asthma in 2014 based on state rates (BRFSS) applied to population estimates (U.S. Census).
[6] Adding across rows does not produce valid estimates. Adding the disease categories (asthma, COPD, etc.) will double-count people who have been diagnosed with more than one disease.
[7] COPD estimates are for adults 18 and over who have been diagnosed within their lifetime, based on state rates (BRFSS) applied to population estimates (U.S. Census).
[8] CV disease is cardiovascular disease and estimates are for adults 18 and over who have been diagnosed within their lifetime, based on state rates (BRFSS) applied to population estimates (U.S. Census).
[9] Diabetes estimates are for adults 18 and over who have been diagnosed within their lifetime, based on state rates (BRFSS) applied to population estimates (U.S. Census).
[10] Poverty estimates come from the U.S. Census Bureau and are for all ages.

## Change in Food Security

Food yields will be affected by climate change. It will affect populations around the world but, primarily, climate change will make a greater impact in regions where warmer temperatures and reduced rainfall will likely occur. Coastal cities will be affected by a reduction in food supply caused by rising sea levels, heavy storms and floods, and a reduction of arable land. These negative factors are increased by acidification, warmer waters, and reduced river flows (Projected Climate Change and Its Impacts, 2007).

## Climate Change and Infectious Diseases

The distribution and incidence of infectious diseases due to climate change effects is projected to increase globally. This fact implies crises with economic and social implications. Malaria and other vector-transmitted diseases get most of the attention from climate-change scientists and health professionals. Lafferty (2009) suggested that latitudinal, altitudinal, seasonal, and interannual associations between climate and disease along with historical and experimental evidence indicate climate, along with many other factors, can affect infectious diseases in a nonlinear fashion.

### Malaria and Climate Change

Climate change will alter the patterns and spread of malaria transmission. Rainfall acts not only on the persistence of water bodies but also on physical and biochemical characteristics of aquatic environments. Heavy rains and flooding are known to cause major malaria outbreaks in semiarid or arid lowlands; spatial and temporal variations in rainfall determine the nature and scale of malaria transmission in highland areas (Gilioli and Mariani, 2011).

After El Niño events in South America and Asia, malaria transmission has been common, which indicates the strong association between higher temperatures and increased rainfall that creates the right environment for an increase in mosquito breeding sites because of surface water collections. Climate variability and drastic weather events have shown to have a direct impact on the increased risk of a St. Louis encephalitis (SLE) outbreak. In general, SLE is restricted to specific areas, such as south of the 20°C June isotherm, but it has spread northerly where temperatures have been warmer in recent years. SLE also has the tendency to spread in periods of hot weather when temperatures are greater than 30°C for seven consecutive days (Hunter, 2003).

## Endemic Chagas Disease

Chagas disease (American trypanosomiasis) is a vector-borne disease. The disease is caused by a protozoon (*Trypanosoma cruzi*); the transmission of the disease occurs through various species of "kissing bugs" (Triatominae) via bites, blood transfusion, and other means. The disease affects more than 8 million people in Latin America. In the United States, there are 300,187 individuals infected by Chagas disease. Spain has the second highest number with an estimated 47,738–67,423 individuals (Rassi et al., 2010). The disease symptoms develop gradually, principally affecting the heart, where patients show progressive heart failure, cardiac arrhythmias, or both, as well as intestinal problems (Aagaard-Hansen and Chaignat, 2010).

Chagas disease is the leading cause of the death and disability-adjusted life years (DALYs) lost resulting from neglected tropical diseases (NTDs) in the Latin American and Caribbean (LAC) region. Chagas disease is one of the most prevalent NTDs in the Americas, and it is among the primary causes of annual deaths and DALYs lost (Hotez et al., 2012).

## Climate Change and Chagas Disease

Chagas disease is one of the most significant climate-sensitive vector-borne diseases in South America, and it is spreading throughout the continent. The globalization of the disease through climate change as well as other factors, such as migration (it can be increased by climate change impacts in certain regions) and blood transfusion, is a concern for developed countries. These aspects are modifying and accelerating the transmission rate and distribution of the disease. The relation between Chagas disease and changes in temperature has been studied basically since the beginning of the disease's parasitological, clinical, and epidemiological description. In 1916, Arthur Neiva, a Brazilian scientist and politician, reported the first cases of Chagas disease in Potosi, Bolivia. He conducted experimental work on the relation linking the influence of temperature variability on the evolution of *Triatoma infestans* embryos. The conclusion of the experiment demonstrated that warming accelerates the embryonic period (Carcavallo, 1999). Therefore, temperature can affect the distribution of the triatomine bug and the pathogen transmission efficacy through the vector. Changes in temperature and its relation to the transmission risk of vector-borne diseases are as follows:

- Increase or decrease in survival of vector
- Changes in rate of vector population growth
- Changes in vector feeding behavior
- Changes in susceptibility of vector to pathogens

- Changes in incubation period of pathogen
- Changes in seasonality of vector activity
- Changes in seasonality of pathogen transmission (Hunter, 2003)

Figure 11.2 shows there are two phases of the human disease: the acute, which will start about one week after the initial infection and is habitually asymptomatic; and the chronic, which is subdivided into indeterminate and clinical (cardiac, digestive, or mixed) forms. The thickness of arrows generally point to the relative probability of a depicted pathway (Rassi et al., 2007).

Theoretically, climate change could affect the transmission rate of Chagas disease, but so far evidence to support this theory is insufficient. Chagas is a climate-sensitive disease that responds to temperatures, but other important factors are involved in the spread of this disease and others like it (Haines, 2012). Nonetheless, researchers consider it necessary to emphasize the susceptibility of Chagas disease to temperatures and the "globalization" of the disease to nonendemic countries. An estimated 300,000 immigrants in the

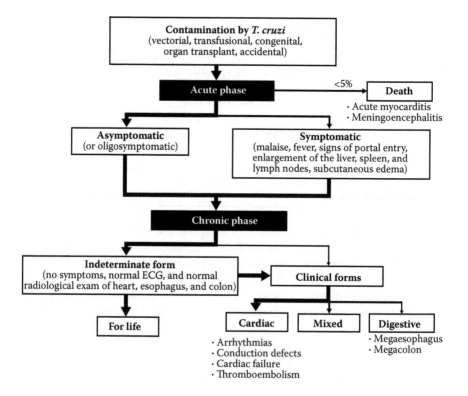

**FIGURE 11.2**
Diagram of the natural history of chagas disease. (From Rassi, A. et al., *Circulation*, 115(9) 1101–1108, 2007.)

United States are chronically infected with *T. cruzi*, with a projected yearly estimate of 30,000 to 45,000 cases of Chagasic cardiomyopathy, and as many as 300 congenital infections (Klein et al., 2012).

## Health as Central to Sustainable Development

The sustainable development of society is related to the well-being of populations. Their physical, social, and psychological good health is important to maintain a balanced and productive society. Climate change directly harms human beings in extreme weather events and indirectly with changes in water, air, food quality and quantity, ecosystems, agriculture, livelihoods, and infrastructure (Confalonieri et al., 2007a).

Population health has improved greatly over the last 50 years. For example, the average life expectancy at birth has increased worldwide since the 1950s, first in the developed world (World Bank Group, 2012c). In the 21st century, however, some developed countries will experience a decline in life expectancy according to the *New England Journal of Medicine*. S. Jay Olshansky, PhD, of the University of Illinois at Chicago, explains that obesity will have a negative effect on life expectancy. In the United States, life expectancy could drop by 5 years in the next few decades unless obesity is combated aggressively to slow the rate of this chronic medical condition (Dollemore, 2005).

Inequalities in public health exist throughout different regions of the world. In many countries in Africa, life expectancy has dropped in the last 20 years mainly because of the HIV/AIDS epidemic, and some populations in the African continent are currently experiencing a 20% increase in infected inhabitants (e.g., Sub-Sahara Africa, which is disproportionally affected) (United Nations Programme on HIV/AIDS, 2011).

The United Nations Millennium Development Goal (MDG) of reducing under-5 mortality rates by two-thirds by 2015 is unlikely to be reached (Sachs and McArthur, 2005). As stated earlier, populations living in poverty throughout the developing world are deeply burdened with untreated communicable diseases and are habitually marginalized by the health sector. These diseases are currently referred to as *neglected diseases of neglected populations*. The neglected diseases create social and financial burdens to the individual, the family, the community, and the nation (e.g., Chagas disease, ectoparasitic skin infestations, and parasitic zoonoses) (Ehrenberg and Ault, 2005). These health issues are an indication of how people are disproportionally affected by diseases; it provides an idea about how climate change can exacerbate the health risks to deprived populations that are exposed throughout the globe.

Sustainable development in megacities and health has a mutual-beneficial relationship that conceptualizes what an all-inclusive and equitable development seeks to achieve. As health is a contributor to development, it is an integral part of human well-being and improves a megacity's economic output, labor productivity, and social stability, and enhances the environmental security element (United Nations General Assembly Open Working Group on Sustainable Development Goals, 2014). These elements are interrelated and interdependent; investment in health and ill-health prevention is significant to reduce vulnerabilities of populations' health risks and create the path to sustainable development.

## Population Health Sensitivity to Climate Change

Assessing climate change impacts on health and the vulnerability of the urban poor requires an understanding of the population's capacity to confront and mitigate the impacts of climate change, and the sensitivity of the population to health risks associated with climate variability (Ebi et al., 2005). The effects of climate change on health based on five empirical studies can be summarized as follows (Parry et al., 2007a):

- Health impacts of individual extreme events, for example, heat waves, floods, storms, droughts, and extreme cold
- Spatial studies where climate is an explanatory variable in the distribution of the disease or the disease vector
- Temporal studies assessing the health effects of interannual climate variability, of short-term (daily, weekly) changes in temperature or rainfall, and of longer-term (decadal) changes in the context of detecting early effects of climate change
- Experimental laboratory and field studies of vector, pathogen, or plant (allergen) biology
- Intervention studies that investigate the effectiveness of public-health measures to protect people from climate hazards

These studies conducted on climate change effects on health are particularly challenging due to several factors, such as the environmental conditions of the living areas, and the adaptive behavior of people (Portier et al., 2010a). A variety of different factors determine the vulnerability of the population to climate change effects on health including biological susceptibility, the population's social and financial well-being, and their built environment (Portier et al., 2010).

## Preventive Measures and Health Risks
## Associated with Climate Change

The World Health Organization created the Consultation on the Essential Public Health Package to Enhance Climate Change Resilience to develop adequate preventive measures to mitigate climate change effects and to determine the scale to which health risks are already incorporated into climate change adaptation planning (Neira et al., 2010).

The concerns related to climate change and the health impacts on populations are injuries, disability, drowning, and heat and cold stress. There are other health risks associated with climate change with an indirect impact, such as water- and food-borne diseases, malnutrition, psychological stress, and vector-borne diseases (e.g., malaria, dengue, kala-azar, plague, Nipah virus, swine flu, bird flu, SARS, and chikungunya). The mission of the Climate Change and Health Promotion Unit of the World Health Organization is to build the capacity of health systems networks to confront the health risks associated with climate change (Neira et al., 2010).

Extreme weather events have been increasingly frequent in the last two decades and climate change effects on human health are presently evident, as a growing number of studies led by many international organizations, such as the Intergovernmental Panel on Climate Change, demonstrate the effects of climate change on the spread of infectious diseases (Haines et al., 2006).

## Vulnerability of Populations

The effects of climate change on health have been reported by the Intergovernmental Panel on Climate Change (Confalonieri et al., 2007b). Direct impacts such as thermal stress, floods, and storms; and indirect impacts such as borne-vector diseases, water-borne pathogens, water quality, air quality, and food availability and quality are concerns for urban populations. The socioeconomic situations of urban residents as well as the environmental conditions of the city are important factors that amplify the impacts of climate change on population's health (WHO, 2003).

Another factor to assess the vulnerability of populations is the existing city's adaptation and mitigation strategies and the viability of the institutions, technology in place, and risk management planning to be implemented when the city's leaders need to act upon the situation (WHO, 2003). The health impacts that are relatively direct are usually caused by weather extremes, in addition to the health consequences of various processes of environmental change and ecological disruption that occur in response to climate change. Furthermore, there are diverse health consequences that need to be accounted for (e.g., traumatic, infectious, nutritional, psychological) that occur in demoralized and displaced populations in the wake of climate-induced economic dislocation, environmental decline, and conflict situations (McMichael et al., 2003a).

## Health and Poverty

Poverty is an important factor that adds to the health effects of climate change on populations. The insufficient access to health care for the poor is detrimental to the poor's susceptibility to drastic weather events. The risk of displacement is another factor of climate change for the poor, and the risk of disease is much higher. Therefore, preexisting conditions and situations increase the stress and vulnerability of the poor to extreme weather events and environmental degradation (Portier et al., 2010).

Major cities in the developing world are experiencing an increased number of people living in slums. A particular characteristic of big cities in the developing world in comparison with cities in industrialized countries is that urban growth is generally uncontrolled.

## Urban Health Challenges

The urban environments of densely populated cities are already exposed to health-threatening conditions. The cost of major environmental changes due to a growing population in megacities is closely related to urbanization and, therefore, an increase in public health risks.

Massive urban growth is also accompanied by urban poverty. The reasons include the absence of urban planning and regulatory institutions. These gaps needed to be addressed. Developing countries are experiencing a fast and extended urbanization as megacities are becoming common in low- and middle-income countries (Frenk and Gomez-Dantes, 2010). In many developing countries, the proportion of urban poor is increasing faster than the overall rate of urban population growth. As a result, the capacity of cities to provide the necessary services is being challenged by rapid urban growth (Cohen, 2006).

Health services are becoming increasingly complex as the lack of resources increases the pressure to provide efficient care to the urban poor. This lack is implicit in the term *underdevelopment* and in what French sociologist Alain Touraine called *maldevelopment*. This term carries a qualitative notion that refers to a discrepancy between the needs of a specific population and the responses generated (Frenk and Gomez-Dantes, 2010).

Maldevelopment is visible in megacities due to poor urban planning and a lack of adequate policies that fit the needs of the population. These urban environments present new problems that coexist with old ones as the public health risks and the inequalities increase (Frenk and Gomez-Dantes, 2010). Therefore, "accumulation by dispossession" (a concept presented by Marxist geographer David Harvey) becomes a common denominator in the health care system in developing countries when dealing with the urban poor. Very few resources are available and distributed to the impoverished, so they lack adequate services to manage health risks. Global environmental problems place added pressures on poor urban centers. Thus, the growth of slums and climate change impacts present a new challenge that the health sector needs to act upon. For example, dengue fever, which was an epidemic illness at the beginning of the 20th century, is becoming an endemic urban disease associated with those problems (Frenk and Gomez-Dantes, 2010). As prevention is the core of public health, the impact of climate change in the urban poor population can be minimized through early action and deterrent measures.

## Vulnerable Urban Populations

Disease burdens can drastically increase as climate change increases temperatures in urban areas. Major cities are experiencing a higher rate of urban population growth than rural areas and that, along with the increased percentage of low income residents that are not capable of mitigating the effects of drastic weather events, magnifies the vulnerability of the urban poor to climate change.

The United Nations conducted a comprehensive study on urbanization, and it concluded that the urban population increased from 220 million in 1900 to 732 million in 1950 and is estimated to have reached 3.2 billion in 2005 (United Nations Department of Economic and Social Affairs/Population Division, 2012). Major cities in developing countries are experiencing a sharp increase in urban populations in comparison to industrialized countries. The urban population was comprised of 74% of developed regions and 43% of developing regions. By 2030, approximately 60% of the global population will be living in urban areas (Parry et al., 2007b).

Slums in developing countries help increase the risk of climate-change-related impacts, because population density plays a major factor in exacerbating the issues faced by the urban population. The frequency and intensity of heat waves will grow as higher temperatures and the heat-island effect work synergistically in megacities (Parry et al., 2007b). Slums and precarious settlements are a tangible reality in urban areas that show deprivation and exclusion among the urban poor. Therefore, all these inequalities and factors shape this particular segment of society and expose vulnerabilities to climate change, which can be in the world's most life-threatening environments (Bartlett et al., 2012).

---

## Climate Change and Urban Children: The Forgotten Population

Children with physical, physiologic, and cognitive immaturity are most often susceptible to health risks and subsequently climate-change effects. Socioeconomic factors also play an important role in adaptation and preparedness to climate variability and health impacts on children living in urban areas. As drastic weather events gradually increase, children will disproportionally suffer the consequences of climate variability and, as a result, they will be particularly exposed to certain infectious diseases, air pollution, and stress from heat waves (Shea, 2007).

There is not enough data to compare mortality and illness rates among children in developed and developing countries. One study compared heat-wave-related deaths in three major cities—Delhi, Sao Paulo, and London—between 1991 and 1994. The cities in the study are amalgamated into different demographic groups composed of diverse ages and socioeconomic statuses. Children were associated with the majority of deaths. Most of the fatalities occurred under the age of 15 due to communicable diseases. In contrast, London and Sao Paulo presented a different scenario where the population over 65 experienced 80% and 48% of deaths, respectively. Common data reflected that over 50% of all deaths were attributed to cardio-respiratory causes. According to the study, children need particular attention during heat waves because of their vulnerability to climate change and inability to adapt to climate hazards (Sverdlik, 2011).

More than one billion children are living in urban areas and the number is increasing rapidly, as half of all people already live in major urban settlements. Urban children are marginalized due to lack of access to health care and facilities to prevent and treat diseases and disaster risks (Bartlett et al., 2012).

### Children's Health at Risk

New data and climate models continue to surface showing greenhouse gas (GHG) emissions modifying the atmosphere and resulting in higher mean temperatures. The climate variability created by GHG emissions creates severe storms and droughts leading to serious implications for children's health. The health impacts of climate change in children will be equally felt in developed and developing nations. Most of the mortality and morbidity rates related to climate change will come from the accessibility of drinkable water, shortage of food, and the accelerated spread of vector-borne diseases of which children are highly susceptible (Bernstein and Myers, 2011).

Climate variability (e.g., higher temperatures, heavy rains, and droughts) leads to outbreaks of diarrheal disease, including cholera, cryptosporidiosis, and other water-borne diseases. More than one billion people are presently struggling to have access to drinkable water, food, and the basic resources for subsistence.

Diarrheal diseases are the cause of approximately 7% of childhood deaths worldwide; unsafe water and poor sanitation and hygiene are the main factors for the illnesses (Bernstein and Myers, 2011).

### Climate Change and Undernourishment

Quantity and quality of nutritional food for children represent perhaps the greatest threat to their health. Again, more than one billion people suffer from malnutrition, and climate change will have a drastic impact on the accessibility of food and its nutritional value. Considering the statistics from undernourished children in developing countries, nearly one in three children are underdeveloped, showing the correlation between undernourishment and childhood morbidity and mortality (Bernstein and Myers, 2011).

### Air Pollution and Urban Children

Children living in urban areas are susceptible to air pollution. Asthma in children causes school absenteeism and, more important, is the primary cause of morbidity. Air pollution triggers asthma attacks, and climate change is increasing the effects of pollution and air particulates that directly affect children's health (Bernstein and Myers, 2011).

### Scarcity of Resources and Displacement of Children

In certain regions of the world, climate change makes food and water supplies scarce. Cities near coastlines will become unlivable and will make population displacement a common path to chaos where children will suffer the most. Approximately one-third of the world population lives within 100 kilometers of the shore and less than 76.2 centimeters above sea level, and climate change impacts on urban populations will create a vast number of refugees, which will affect megacities' social and economic activity. Mass population displacement increases risk and creates conflict over natural resources. The population displacement will affect the health status of people in general, in particular, children being the population's sector that will suffer the most severe physical trauma, malnutrition, and infectious and psychiatric disease (Bernstein and Myers, 2011).

## Health of the Homeless and Climate Change

Homeless populations have greater exposure and propensity to acquire diseases than the housed population. As climate change effects are more frequent, the homeless have less protection from climate events and, therefore, have a higher vulnerability to climate-sensitive diseases. The poor tend to occupy high-risk urban areas and this results in greater rates of disease and death due to the health effects of heat waves, air pollution, droughts, floods, and vector-borne diseases consequential to drastic climate variability (Ramin and Svoboda, 2009).

Urban areas are experiencing not only a steady growth in population but an increasing homeless population that is marginalized by social and economic factors, despite the positive outlook for capital growth and prosperity in the 21st century among megacities. As of today, there is not enough data representing the effects of climate change on the homeless population (Ramin and Svoboda, 2009).

Knowing this segment of society is highly vulnerable to climate change, because of their inability to adapt and their lack of access to health care, the homeless are likely to be highly susceptible to climate change. In New York City, with its population of approximately 8.5 million, nearly 60,000 homeless men, women, and children are housed in shelters and at least 3,100 sleep on the streets in winter. Approximately 1% of individuals who have experienced homelessness in the last 5 years have used a New York Center shelter annually (Ramin and Svoboda, 2009).

Chronic diseases, heart disease, and cardiovascular disease as well as hypertension, high cholesterol, and diabetes are health risks common in the urban poor and, subsequently, the homeless population living in cities. Climate change and its impacts on health exacerbate those health conditions and accelerate the rate of morbidity and mortality among them. Homeless children in the United States are experiencing an increase in asthma attacks (six times the national rate) and a frequent occurrence of mental illness such as depression and schizophrenia.

Climate variability and the inaccessibility of health centers and adequate health care for the poor affects homeless children, especially since children who live on the streets are exposed to warmer temperatures that create a fertile ground for the spread of diseases. Low socioeconomic status and poor housing conditions are known risk factors for vulnerability to the effects of climate change on health and ultimately cause death among this particular and deprived urban population (Ramin and Svoboda, 2009).

## Incidence, Prevalence, and Morbidity

The prevalence of asthma has increased since the early 1960s and this trend affects children and adults with diverse backgrounds, lifestyles, and locations. Furthermore, in a global context, the cases of other atopic disorders have also risen during the same time frame. These disorders are allergic rhinitis, atopic eczema, and urticaria (Beggs and Bambrick, 2006).

The connection between allergic diseases such as asthma and aeroallergens is well established. Asthma incidents have increased as a result of faster plant growth, earlier plant maturity, and longer growing seasons, plus earlier pollen seasons, increased season duration, and increases in both pollen quantity and allergenicity with consequences leading to a rise in disease frequency. Climate change can be a potential factor for the increase of susceptibility of asthma and also its morbidity within the affected population's group. Infants are highly vulnerable and their exposure to allergens induces susceptibility to asthma and other atopic conditions such as eczema and allergic rhinitis (Beggs and Bambrick, 2006).

The substantial increase in pollen production as a result of elevated $CO_2$ concentrations is a cause of concern, and the trends of increased pollen amounts in the late 1990s are associated with local rises in temperatures. Consequently, the evidence suggests increased temperatures are connected to the strong allergenicity in pollen from trees (Beggs and Bambrick, 2006).

Pollen production from ragweed grown in chambers at the carbon dioxide concentration of a century ago (about 280 parts per million [ppm]) was about 5 grams per plant. At today's approximate carbon dioxide level, it is about 10 grams; and at a level projected to occur in about 2075 under the higher emissions scenario, it will be approximately 20 grams (Karl et al., 2009).

## Climate Change, Food Security, and Health Impacts

Food security will be at risk even in developed countries as anthropogenic GHG emissions and natural "climate forcings" (other mechanisms that lead to climate variability such as stratospheric volcanic aerosols) will have detrimental impacts on agriculture and food processing. Responding to these risks and adapting to climate change becomes crucial in the agricultural and food production industries, and the focus of the response will be mitigation strategies that propose modifying farming and food systems to reduce GHG emissions associated with the food chain. The use of biofuels can provide another strategy to mitigate climate change impacts on agriculture and food production. All of the aforementioned initiatives will change the type of food availability to consumers in developed countries. Many nutritional components of the food consumed will change, and consumer choice will be influenced by these modifications (Lake et al., 2012).

## Food Sourcing, Consumption, and Its Consequences

The modification of food belts is projected with climate change, and food consumed will be distributed from different regions around the globe. For example, according to the World Cancer Research Fund/American Institute for Cancer Research of 2007, the United Kingdom obtains much of its selenium from grain (the element selenium may be protective against several types of cancer). From 1970 to 2000, there was a 50% reduction in dietary selenium consumption in the United Kingdom, which was the result of a shift to the consumption of grain that was imported from Canada. Food safety risks are also of importance. Climate change alters the locations where food is produced. As human behavior is affected by climate, and with the impacts of climate change, people may tend to consume different foods, and subsequently this will impact food safety and nutrition (Lake et al., 2012).

Without a global commitment to reducing GHG emissions from all sectors, including agriculture, no amount of agricultural adaptation will be sufficient under the destabilized climate of the future (Beddington et al., 2012).

Globally, food demand will grow in the future as the population grows and changes diets, and food supply must somewhat exceed food demand if everyone's needs are to be met and food prices are to remain affordable. Under a "business as usual" approach, food production will decrease over time due to land degradation, climate change, and the emergence of new pests (Beddington et al., 2012).

The large resulting gap between food supply and demand can be bridged by simultaneously applying three general approaches:

1. Avoiding losses in current productive capacity can include actions to adapt to or mitigate climate change, to reduce land and water degradation, and to protect against emerging pests and disease.

2. Increasing agricultural production per unit land area can be achieved through use of improved technologies, practices, and policies; more efficient use of existing agricultural land; and targeted expansion of agricultural land and water use (where negative environmental impacts are minimal).

3. Reducing food demand can be accomplished through efforts to promote healthier and more sustainable food choices, and to reduce food waste across supply chains.

None of these three approaches alone is sufficient, and all three require substantial innovation in the food system (Beddington et al., 2012).

## Assessment of the Potential Future Health Impacts of Climate Change

Developing an evaluation of potential future effects of climate change on health can provide a valuable tool to forecast and address these risks and consequences associated with climate variability. The framework for the assessment has to highlight issues presented by climate change according to the region, health outcomes, and adaptation. The challenges and limitations faced by developing a comprehensive evaluation include the following issues (Confalonieri et al., 2007b):

- Limited region-specific projections of changes in exposure of importance to human health
- The consideration of multiple, interacting, and multicausal health outcomes
- The difficulty of attributing health outcomes to climate or climate change per se
- The difficulty of generalizing health outcomes from one setting to another, when many diseases (such as malaria) have important local transmission dynamics that cannot easily be represented in simple relationships; limited inclusion of different developmental scenarios in health projections
- The difficulty in identifying climate-related thresholds for population health
- Limited understanding of the extent, rate, limiting forces, and major drivers of adaptation of human populations to a changing climate

## Disease Outbreak and Climate Change

There is clear evidence of the relation between climate change, health risks, and the long-term consequences to the global population. Disease outbreak in uncontrollable proportions disrupts lives and displaces millions of people, forcing them to migrate. The ability to stop what is happening will continue to decrease if world leaders do not take action. It is nearly impossible to roll back the clock or even slow changes that are already in the system. The increase of diseases that we are witnessing in certain regions is evidence of how climatic change affects the outbreak of disease (McMichael et al., 2003b).

An argument can be presented because there is an urgent need to adapt and subsequently mitigate climate change impacts. Although many major cities around the world are already acting upon climate change effects, an immediate action plan is urgent and it needs to include a comprehensive health-risk assessment. Assessments cannot predict the future; caution needs to guide the process when developing and implementing risk management plans to mitigate climate change effects. Approximations and likely impact indicators of climate change effects are acceptable (McMichael et al., 2003b).

## Projecting Health Impacts of Climate Change

Adaptation strategies that include and implement a risk management plan addressing the health impacts of climate change are essential in order to protect and develop a realistic framework that is feasible to protect the megacities' urban poor populations (Nerlander et al., 2009).

The long-term impacts of climate change need to be addressed in order to develop a comprehensive risk-management framework. Populations at risk are also considered in the projections of climate-change-related health impacts. The economic situation of the population and the region or country analyzed cannot be directly related to disease burdens, because climate change health risks and the gross domestic product are determined by different causes such as climate, and environmental and social factors. The well-being and good health of populations is determined by factors other than income per capita. Therefore, the increased per capita income does not improve the health of populations (Nerlander et al., 2009).

## Projecting Heat-Related Mortality

Many studies have been conducted by the World Health Organization and the United Nations that provided convincing evidence that linked climate change and premature death from many causes. Although these studies are valuable in understanding and mitigating the effects of drastic climate events on health, more efforts are needed to evaluate and eventually predict how climate change affects the health of populations. The average annual temperature in the United States has increased by 1°F during the 20th century. This national standard has been surpassed in the metropolitan area of New York. In the years between 1990 and 1997, New York temperature has increased an average of 2°F. The prior statement associates climate change on ambient temperatures and related mortality (Knowlton et al., 2007).

The increased temperature trend is still in place as warming temperatures and greenhouse gas emissions will continue to affect the climate well through the 21st century. The projections of a warmer climate are astonishing. The recent trends on the impact of anthropogenic emissions on the global climate suggest that annual average temperatures will rise between 2.5°F and 6.5°F, with summer temperature increases of 2.7°F to 7.6°F, by the year 2050 and beyond (Knowlton et al., 2007).

## The Urban Heat Island in Relation to Public Health

Climate and the atmosphere are influenced by human activities through air pollution, GHG emissions, land alteration, and airborne particles. Higher surfaces and near-surface air temperatures are coupled in urban areas, and this creates a hotter atmosphere, in addition to higher energy consumption and increased smog formation. Urban heat islands (UHIs) are created by man-made exteriors, including dark roofs, asphalt lots, and paths, which absorb sunlight and reradiate energy as heat (Rosenzweig et al., 2011a).

The heat island effect leads to increased temperatures that affect public health and creates a high demand for cooling energy in commercial and residential buildings in summer, which increases GHG emissions. Urban populations are exposed to particulate matter, carbon monoxide, sulfur dioxide, and nitrogen oxides that can exacerbate respiratory problems and damage lung tissue (Rosenzweig et al., 2011b).

---

# Mental Health Effects of Natural Disasters in Times of Climate Change

*Ariel Durosky, Anne Hilburn, Ari Lowell, and Yuval Neria*

Consequences of climate change present numerous unprecedented challenges to the global community, particularly from the rise of weather-related natural disasters resulting from extreme temperatures (Houghton et al., 2001). In addition to the physical damage caused by such extreme weather disasters, the mental health impact of such events on victims is considerable. Many individuals illustrate resilience, including those who may exhibit symptomatology consistent with posttraumatic stress disorder (PTSD) directly in the aftermath, by developing coping mechanisms, often recovering without the need for clinical treatment (Neria, Nandi, and Galea, 2008; Norris et al., 2002). However, a significant group develops long-term mental health effects that may include PTSD, depression, adjustment disorders, anxiety, and traumatic grief (Hobfoll et al., 2007). Among those, PTSD is the most common malady resulting from exposure to a disaster and is, therefore, the most common subject of disaster research (Neria and Shultz, 2012). The prevalence of reported PTSD among those who have been exposed to hazards during a natural disaster can reach up to 60%, and up to 57% for overall mental health morbidity in postdisaster populations (Udomratn, 2008; Young, Ford, Friedman, and Gusman, 2008).

The following explores the impact of extreme weather events on mental health, particularly within the context of human-induced change. We also outline the trauma signature (TSIG) analysis, a system developed to improve the manner in which disaster recovery efforts address possible psychological outcomes. Future directions are discussed.

Psychological repercussions of disasters vary widely, as each event generates a unique combination of exposures to risks and hazards. This difference is illustrated by comparing the 2010 Haiti earthquake and the 2014 South Napa earthquake. Though the two disasters share exposure similarities, such as perceived threat to life, economic loss, and personal injury, significant differences can be found, which may affect mental health outcomes and ultimately treatment responses. Some defining features of the Haiti disaster include loss of loved ones, witnessed death, and total destruction of property (Shultz, Marcelin, Madanes, Espinel, and Neria, 2011). Comparatively, the South Napa earthquake presented little to no loss of life or witnessed death, and although property damage was significant, total destruction was minor (Henn, Smorodinsky, Attfield, and Dobson, 2015). This example demonstrates the individuality of each disaster.

In addition to symptom variability related to disaster type, Shultz and colleagues (2011), among others, identified the severity of exposure as the key determinant of adverse mental health effects. Individuals directly exposed to the event through injuries related to the disaster or witnessing the deaths of others may be at a greater risk of developing long-term mental health effects than those indirectly exposed, such as through the death of a loved one. The authors determined that personal injury and perceived threat to life are among the greatest predictors of long-term psychological impairment. Other significant risk factors for mental health problems include loss of life of a significant other, displacement and relocation, loss of property, and financial instability (Neria and Shultz, 2012). Moreover, research conducted after the 2004 tsunami in Sri Lanka has indicated that lack of fulfillment of basic needs, such as food and water, separation from loved ones, and total destruction of property were significant predictors of both PTSD and depression (Dewaraja and Kawamura, 2006). Geographic location is also noteworthy in relation to disaster impact, as research has shown close proximity to the disaster zone is associated with PTSD across different forms of trauma (Shultz et al., 2011).

In addition to direct exposure, other specific factors can help identify subgroups with increased risk for long-term mental health effect. These subgroups include individuals with poor social support, children, the elderly, women, ethnic minorities, and those with a prior history of mental illness (Math, Nirmala, Moirangthem, and Kumar, 2015). These risk factors should be taken into account when assessing risk as well as providing outreach and services.

As discussed earlier, the adverse impact of extreme weather disasters on mental health may be considerable. The case analysis of Haiti earthquake is a prime example of the disconnection between the recognition for the significant impact of mental health and appropriate, evidence-based mental health interventions (Shultz et al., 2011). Extreme levels of several major risk factors, such as exposure to personal physical injury and witnessed physical injury, loss of loved ones, witnessed death, loss of resources, and displacement were reported in the wake of this disaster, presumably raising alarm among mental health professionals (Shultz et al., 2011). Despite these known risk levels, however, mental health treatment was not included in most disaster recovery plans (Shultz et al., 2011). Of the little mental health response provided, efforts were uncoordinated, culturally inappropriate, and unsupported by treatment research. Additionally, there was little to no program evaluation to assess efficacy (Shultz and Neria, 2013). Unfortunately, the shortfalls of this case study are not uncommon, exposing the need for a more standardized system that maintains consistency and cohesion, while allowing for geographic and cultural versatility (Hobfoll et al., 2007).

The TSIG analysis model was developed to address such concerns by providing direction and guidance to service providers with inadequate mental health resources and psychosocial support that can be adapted and tailored to the unique conditions of individual disasters (Shultz and Neria, 2013). An adequate mental health psychosocial support (MHPSS) response with the inclusion of the TSIG analysis follows six sequential steps:

1. Trauma signature analysis
2. TSIG targeted preparations of resources
3. Execution and evaluation of evidence-based early interventions
4. Application of validated postdisaster mental health assessments
5. Evidence-based, culturally adapted interventions for high risk affected populations
6. Continued supervision and assessment throughout recovery (Shultz and Neria, 2013)

Steps 1 and 2 are specific to the TSIG model and provide a framework that is essential to identifying the type and severity of a disaster and possible exposures in order to generate and initiate the appropriately tailored mental health response. This disaster analysis consists of developing a hazard profile, creating a unique stressor–risk factor matrix, and establishing a trauma signature summary (Shultz and Neria, 2013). In the direct aftermath of a disaster, initial situational reports from government and nongovernmental organizations can be reviewed as part of efforts to collect data specific to the present disaster. Such reports can contribute information regarding characterizations of the affected community; physical impacts; and statistics on deaths, injuries, the displaced and others affected to detail the hazard profile (Shultz and Neria, 2013). Once the hazard profile has been completed, a literature review can be conducted in order to identify psychological risk factors that match the exposure to hazards previously established in the profile. This information can be outlined in a risk factor matrix during both impact and postimpact, identifying the stressors of hazards, loss, and change as a result of the physical forces of harm. Finally, the initial TSIG analysis is completed with a trauma signature summary composed of major psychological risk factors categorized by exposure to hazards and loss and change from the preceding risk factor matrix, with the addition of severity for each risk factor.

Upon completion, the trauma signature summary should be used to guide the preparation and training of disaster responders, and organize and allocate resources based on identified exposures and risk factors (Step 2) (Shultz, Forbes et al., 2013). Early interventions (Step 3) should focus on promoting a sense of safety, community, and hope for the future (Shultz et al., 2011). Research has shown that individuals who regain a sense of safety are at a substantially lower risk of developing PTSD than those who do not (Bleich, Gelkopf, and Solomon, 2003; Grieger, Fullerton, and Ursano, 2003). Efforts should be made to ensure the validity and efficacy of early interventions. Next, validated assessments must be conducted and those at risk identified, keeping in mind defining exposures of the disaster determined in the trauma signature summary, in addition to specific risk factors mentioned earlier (Step 4). Following assessments and interventions provided should be evidence based and culturally competent (Step 5) (Shultz and Neria, 2013). Mental health care must be tailored to the specific needs of the population in relation to the disaster's trauma signature. Last, continual follow-up assessment of response effectiveness should be maintained throughout the recovery process (Step 6) (Shultz and Neria, 2013). These procedures allow for event-specific analysis of each disaster, allowing for the optimization of the MHPSS response.

Initial TSIG analysis and the subsequent steps for MHPSS responses have been evaluated with a number of disasters, both natural and man-made (Shultz et al., 2011; Shultz and Neria, 2013; Shultz, Walsh, Garfin, Wilson, and Neria, 2014; Shultz, Forbes et al., 2013; Shultz, Garfin et al., 2014; Shultz, McLean et al., 2013). The Great East Japan Disaster is a case study that illustrates a well-executed and excellent mental health response. In March 2011, Japan experienced a massive undersea earthquake, initiating a tsunami that submerged much of the east coast of Honshu and damaged nuclear power plants, resulting in multiple reactor meltdowns. Three days postimpact, a disaster-response committee was established, which mobilized a center for coordination of mental health efforts and resources. Additionally, an informational website was launched to construct national protocols (Shultz, Forbes et al., 2013). This unification of management allowed responders to coordinate early evidence-based interventions across affected populations in a timely manner, demonstrating proper implementation of Steps 2 and 3 of the MHPSS response. Japan's mental health response addressed vital issues including identification and outreach to survivors and disaster responders, and as well as high risk and remote individuals (Steps 4 and 5). Specific responses were comprised of both psychoeducation and the promotion of validated, culturally appropriate interventions. It is important to note that, in compliance with the TSIG model, Japan's disaster recovery only included evidence-based interventions (Shultz, Forbes et al., 2013).

In summary, with the increased likelihood and frequency of natural disasters and other catastrophic events due to climate change, it is important that efforts are made to improve preparedness for mass-scale mental health challenges. Implementing the TSIG system as a universal standard could help to ensure that mental health needs are promptly met while simultaneously minimizing the residual damages on those affected. The TSIG analysis is continually undergoing development and refinement, as well as seeking opportunities for validation. Ultimately, once finalized, TSIG is intended to provide a scientific basis to aid in MHPSS response decisions in the wake of disasters.

# 12

# Climate Change Impacts on Health: Urban Poor and Air Pollution in European Context

**Maria Cristina Mammarella, Vincenzo Costigliola, and Giovanni Grandoni**

## Introduction

The history of socioeconomic and political development in modern society has been closely intertwined with the human ability to transform and use energy in all its forms and make it accessible both widely and efficiently. The use and the possible transformation of energy have a substantial impact on society, the economy, and natural environments. Climate change further complicates the energy challenge.

Tackling climate change at an international level requires programs with accurate information for policy-makers. It is critical that resources are made available for research, information, and training that should be delivered by a qualified international task force. It is also important to develop equipment, strategies, and preventive measures while creating infrastructures and networks at both the national and international levels to respond swiftly and efficiently to emergency situations (Fernando, Klaic et al., 2012).

To reduce the risks caused by climate change, efficient and environmentally sound systems for energy transformation and energy use should be adopted. Energy supports human life at all levels; every human and every nation need access to energy to grow and prosper in a way that is socially and economically sustainable for them and the environment (Pontificium Consilium de Justitia et Pace, 2013).

People are therefore responsible for *how* energy is produced and *how* it is used. As a result, the concept of sustainable development should be expanded and diversified, with the term energy being combined with attributes from anthropology and ethics, such as fair, conscious, and equitable. A new avenue may be indicated to solve problems linked to energy, such as wars, terrorism, food insecurity, pollution, and climate change (Pope Francis, 2015). Researchers, the scientific community, and academia should take into account civil, cultural, and social elements because they are closely connected with scientific and technical concerns. The first steps are being taken.

Major cities are major users of energy in different forms (lighting, transportation, services, food manufacturing, industrial activities, trade, etc.), so much so that energy is a founding and characterizing element of cities. Cities are a key center of energy transformation and use, which, in turn, generates and concentrates pollution, especially air pollution, continually exposing populations to harm. Those who have means or knowledge may try to shun this exposure more or less effectively, whereas the poor will inevitably remain less protected.

## Climate Change and Air Pollution in a European Context

Alongside deforestation and the burning of large quantities of biomasses, much consumption of the energy generated using fossil fuels (oil, coal, and gas) discharges a large amount of carbon dioxide ($CO_2$) into the air. This $CO_2$ enters the carbon cycle between the earth and the atmosphere and has a direct impact on climate, the ecosystems, and biodiversity.

From the beginning of the new industrial age until now, the content of $CO_2$ in the atmosphere has grown continuously. Carbon dioxide is a reference climate-changing gas, since it also interferes with the sea, in particular by reducing its pH. For these reasons, it is closely monitored; in fact, the increase in $CO_2$ has a direct relation with the average increase in the earth's temperature.

To have more certainty and information in this area, the United Nations (UN) and the World Meteorological Organization (WMO) set up the Intergovernmental Panel on Climate Change (IPCC). The aim is to understand the causes of anthropogenic effects on climate, and propose solutions and options for strategies designed to reduce or to adapt to climate change. The ultimate goal is to mitigate adverse impacts on human health and the environment. Harmful effects may be even more devastating in developing countries, due to their geographical position and the living and nutritional habits of their populations, who are closely connected to the land and the climate in which they live. These areas have the highest numbers of destitute and poor people.

Meteorological and greenhouse gas measurement stations have been installed in most areas of the world, so as to have a wide representation of climate diversity. The goal is to monitor meteorological and climate parameters constantly and to assess phenomena on a large scale. Figure 12.1 shows the $CO_2$ results from one monitoring station from 1992 to 2010.

To process the collected data more effectively, these monitoring stations are interlinked with systems of international organizations, including the WMO's Global Atmosphere Watch and Integrated Carbon Observation System (ICOS), presented by the European Commission at COP21 in Paris.

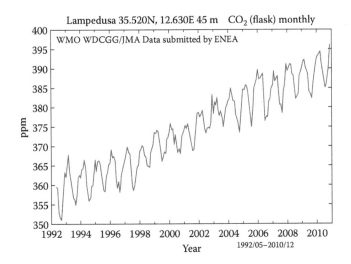

**FIGURE 12.1**
$CO_2$ results from one monitoring station from 1992 to 2010.

Some of these monitoring stations are located in Italy; there is one in Lampedusa, a small island in the heart of the Mediterranean. Due to its particular geographical position, it represents a remote reference station, that is, a station that is not linked to the anthropogenic production of air pollutants, and, therefore, points out a background character for the Mediterranean climate. Lampedusa's climate monitoring station is run by ENEA (Italian National Agency for New Technologies, Energy and Sustainable Economic Development) and has been in operation since 1992 (Artuso et al., 2009).

The data clearly suggests a constant increase of $CO_2$ over time up to the present day despite yearly variations. The daily cycle is much reduced, which confirms that the observation site and its value for background $CO_2$ measurements are representative (Artuso et al., 2009).

At present, the $CO_2$ concentrations in the atmosphere have reached 400 ppm, whereas, at the beginning of the last century, the $CO_2$ concentrations were consistently lower than 300 ppm. If the current energy development pattern is maintained, $CO_2$ concentrations may near the dangerous level of 450 ppm over the course of a few years. This figure is estimated to be the resilience threshold of natural ecosystems and more devastating events. Many scientists see this value as high risk because $CO_2$ cannot exceed 450 ppm of $CO_2$ to stabilize the average temperature increase within 2 degrees centigrade, according to IPCC and Organisation for Economic Co-operation and Development (OECD) estimates (OECD, 2012).

The most visible effects of climate change are already under way, including drought, and rainfall intensity and duration (change in the water regime), as well as the variation of mean temperature on both a local and a regional scale (thermal stress). This last characteristic is well represented by heat waves, which are occurring increasingly often.

As widespread awareness developed of the need to counter the causes of climate change, Europeans have proposed policies and strategies aimed at reducing the consumption of energy from fossil fuels. The intent is to steer human activities toward a more sustainable development model and to mitigate the adverse effects on the environment, thus protecting both human health and the planet. The European Commission focuses on the major cities that should make efforts to improve their strategies for reducing pollution and adapting to climate change through in-depth knowledge and understanding of their specific situation (European Environment Agency, 2012; European Parliament and the Council, 2010).

The European Commission (Covenant of Mayors for Climate and Energy, n.d.) adds that policies, initiatives, and decision-making tools are needed to foster resilience to climate change and that the effects of the decisions taken should be continually monitored. The aim is to predict and prevent threats to human health and biodiversity, due to invasive allochthonous species that may find suitable conditions for their development with increased temperatures.

One avenue to pursue is to undertake initiatives and forms of prevention that are based on the prediction of peaks of air pollutant concentrations and extreme meteorological events, such as heat waves. At the same time, citizens' access to microclimatic and air quality information should be fostered, the development of innovative technological solutions should be increased, and the application of mathematical models should be promoted on a regional and local scale (Mammarella et al., 2012). These actions would make it possible to alert, prepare, and protect the population by minimizing adverse consequences through interventions focused on relieving the effects of exposure for the most vulnerable sections of the population.

To support decision-makers in a typical local setting, such as that of a city, a successful comparison has been made between deterministic mathematical models and neural network stochastic models by ENEA, combined with weather forecast models (Mammarella et al., 2009). Both modeling simulations, applied to the urban area of Phoenix, Arizona, proved suitable because they are specifically designed to predict pollution in cities. The neural networks obtained better results, as they are more suited for predicting acute episodes (Fernando, Mammarella et al., 2012).

Building on experience acquired in years of studies, the ENEA developed an intelligent information system based on neural network models, called A.T.M.O.S.FE.R.A. This system makes it possible to shift the focus of action: steps can be taken before acute pollution episodes take place. Timely interventions can be planned to improve the air quality in urban areas, and the combined effects of extreme meteoclimatic events can be minimized by informing the population models (Mammarella et al., 2005).

The relationship between air pollution and meteorological conditions, specifically the urban planetary boundary layer (PBL) (Grandoni et al., 2010), is a complex nonlinear and dynamic system, since meteorological data represents an exogenous factor in the stochastic model. From a prevention perspective, neural network models are more suitable for reproducing a sudden change in meteorological conditions, as they can represent nonlinear functions, built through an automatic learning process. Therefore, a shift is made from the current emergency-based approach (ex-post) to a scientific and rational prevention-based approach (ex-ante) by adopting measures aimed at averting the negative events that have been predicted.

The neural network system A.T.M.O.S.FE.R.A. makes it possible to predict the hourly air concentrations of pollutants (such as $CO$, $SO_2$, $O_3$, PM10, $NO_2$, and $C_6H_6$), 24, 48, and 72 hours in advance. Air pollution monitoring stations then measure air quality and verify the performance of the neural network system. Historical data sets of measured pollutants, meteorological data sets, and short-term meteorological predictions for specific urban areas help to evaluate pollutant concentrations. A.T.M.O.S.FE.R.A. was developed, performed, and applied to Rome, Milan, and Naples, Italy. Pollution predictive data refers to data collected by the air pollution monitoring stations and represents the portion of territory (usually of the city) in which these stations are located.

## The Impact of Climate Change on Health

From an environmental perspective, the behavior of a large city represents an open system, consisting of the conformation of the urban structure, emission factors due to traffic and winter heating, and meteoclimatic factors, affecting air pollutant concentrations. Air quality is an important indicator of the well-being of citizens.

No one questions the fact that the air we breathe is an essential element of our health; it has been scientifically proven that exposure to mixtures of air pollutants is an important pathogenic factor, especially in diseases of the respiratory system. Since the 1950s, a large number of toxicological and epidemiological studies on the impact of pollution on health have been completed. Never before have we experienced such an exponential progression of respiratory conditions, ranging from asthma to allergies, as in the last decades. Those who suffer the most are the most vulnerable groups: children and the elderly.

Vulnerability means the possibility that, in the presence of specific stimuli, a person may develop a given disease. Environmental factors are often risk factors favoring or triggering the onset of diseases that otherwise would have not developed. The concept of vulnerability does not apply only to people; it can be extended to plants and also to artifacts (imagine the potential impact of pollution on monuments). These topics are the subject of studies and recommendations by European Institutions, which invite member states to take all measures aimed at assessing the air quality and protecting citizens from exposure to potentially hazardous air pollutants.

Asthma and chronic obstructive pulmonary disease (COPD), pulmonary diseases, chronic rhinitis, and respiratory allergies, which are often underestimated and underdiagnosed, are among the most frequent conditions in children and the elderly. They could be dramatically reduced by limiting exposure to environmental risk factors, such as benzene, nitrogen dioxide ($NO_2$), microparticles and nanoparticles, and formaldehyde in the air, both outdoors and indoors. A study has shown that in Europe, 4.6% of deaths from all causes and 31% of inabilities in children from 0 to 4 years of age are attributable to air pollution both outdoors and indoors (Valent et al., 2004).

The respiratory system is the point of entry for toxic agents that, once in the lungs, enter the bloodstream and can attack other organs, leading to diseases from lung cancer to lymphomas, women's breast cancer, leukemia, thyroid diseases, and ischemic heart diseases. The impacts of air pollution on health may be chronic if humans are exposed even to low concentrations for a long time or acute if humans are exposed to high concentrations for a short time. Air pollution continues to be an "invisible killer."

The European Commission has published *The Clean Air Policy Package* (2013), whose aim is to significantly reduce air pollution across the European Union. *The Clean Air Policy Package* consists of a communication on a Clean Air Program for Europe and three legislative proposals on emissions and air pollution. It is estimated that, by 2030, *The Clean Air Policy Package* will "avoid 58,000 premature deaths, save 123,000 km² of ecosystems and 56,000 km² of protected Natura2000 areas from nitrogen pollution, and save 19,000 km² of forest ecosystems from acidification." Air quality standards for the protection of human health are clearly indicated in Directive 2008/50/EC of the European Parliament and of the Council of 21 May 2008 on ambient air quality and cleaner air for Europe. These are also defined in the new national provisions implementing Directive 2004/107/EC of the European Parliament and of the Council of 15 December 2004, relating to arsenic, cadmium, mercury, nickel, and polycyclic aromatic hydrocarbons in ambient air.

In big cities, climate change generates an additional negative impact on human health, as air pollution and the components of the urban structure (cement, steel, asphalt, street canyon effect, urban canopy, etc.) create a synergy with heat waves, which can heighten the negative impacts on people, at least until medium latitudes. During the summer of 2003, Europe was hit by a heat wave (see Figure 12.2) that caused the death of thousands of elderly people, who were not ready for the unusual situation. That heat wave strained welfare systems, administrative structures, and, most especially, first-line physicians.

According to statistical data (European Commission, 2005), the death toll in Europe was as follows:

- About 15,000 in France
- 2,139 in Great Britain

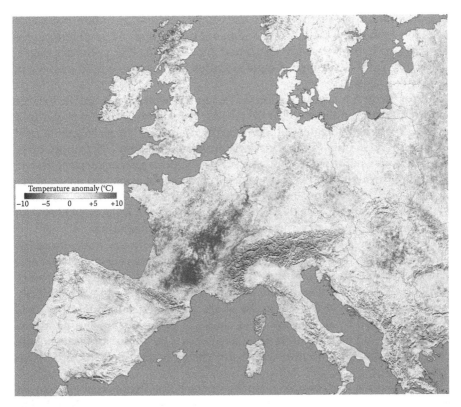

**FIGURE 12.2**
Temperature anomalies in Europe in 2003.

- Between 1,866 and 2,039 in Portugal
- About 300 in Germany
- About 1,500 in the Netherlands

Following this experience, dedicated air-conditioned facilities have been arranged to accommodate people at risk.

## The Urban Poor and European Context

Large cities continue to be seen as a haven of well-being and security for many people, but this prospect often turns out to be a deceit for many who do not succeed in integrating themselves. From a social, economic, and health perspective, they are forced to live on the fringe of society, often in degraded suburbs. Public health and social services are often not easily available to them, so the physical and mental states of these desperate people are exposed to the worst environmental conditions (traffic, pollution, degradation, etc.) and to extreme climate episodes (temperature variations, wind, rain, drought, etc.).

Across Europe, marginalization is a fact that is even more dramatic in large cities and increasingly so due to the current influx of migrants from war-torn countries or economically and climatically distressed areas. Migrants tend to seek refuge mainly in urban areas (European Agenda on Migration, 2015). Thus, it is now a European Union priority to achieve greater social cohesion, including through improved strategies to reduce pollution and adapt to climate change at a national level (European Commission, 2015). Data and surveys made in large cities suggest that marginalization often coincides with an urban fabric that is environmentally and socially degraded.

The presence of poverty turns out to be an extraordinary indicator of the environmental, urban, and social degradation of a city. In peripheral and low-income neighborhoods, these effects are worsened by industrial settlements and legal and illegal dumping sites. Such sites are often strategically located in these areas (Mammarella et al., 2011). Environmental distress is thus heightened by the emissions of chemical pollutants, including asbestos, dioxins, and heavy metals (such as mercury, cadmium and lead, cobalt, chromium, and aluminum), which are certainly the most hazardous to health.

Heavy metals may be taken up by contact (for instance, with colored clothing or toxic products), by ingestion of food and water, or inhaled with air. They penetrate slowly into the body and stratify in different organs and systems, and because of this characteristic, it is extremely difficult to eliminate them (http://medicalpraxis.it/web/, n.d.). Depending on the doses and the frequency of exposure, heavy metals are allergenic, and damage the liver, the kidneys, the nervous system, and the immune system. Exposure to heavy metals has also been considered as a factor promoting the onset of cancer.

These pollutants add to urban pollutants from traffic and heating. The concentrations of $NO_x$, $SO_x$, fine and ultrafine particulate matter, and so on increase dangerously in the streets and alleys of cities into the atmospheric layers closest to ground level because of the strong instability phenomena of the atmosphere due to the low levels of the urban PBL (Mammarella et al., 2012) and to the so-called street canyon effect.

The PBL is an important atmospheric parameter of the meteorological condition because it has an indirect impact on the safety (health) of the population and a direct impact on the security of the environment in which human activities take place (it is one of the causes of fog, of difficulty of the wave transmissions, of insecurity of light aircraft flights, etc.)

(Fernando, Mammarella et al., 2012; Mammarella et al., 2005). With regard to exposure to pollution, mention should be made of the social and health recovery of those who have health problems due to acute or chronic exposure to air pollutants and of associated costs. The recovery of people exposed to air pollutants is perhaps one of the most difficult issues; this is why prevention is always the first option. In addition to essential drug therapies, recovery options have an additional cost and require the patient to move away from the triggering elements to prevent the disease from becoming chronic. If possible, personal, genetic, and family predisposing factors should also be investigated.

A project should be considered for the setting up of "detoxification" centers in places with good air quality (villages and towns in the mountain or by the seaside). In this regard, we would like to recall a program for children from Chernobyl, who have been hosted in a number of European countries for years so as to allow them to breathe less polluted air and eat noncontaminated foods.

## Conclusions

To adequately tackle the problems identified in this chapter that increasingly affect our planet and humanity as a whole, greater awareness is needed at all levels in civil societies and in national and international institutions. Likewise, new development goals should be set intelligently for a sustainable future. Research activities are necessary for developing new technologies that reduce polluting emissions at their source, improve energy efficiency, and reduce food waste, thus saving the planet's resources.

This may all foster a program aimed at monitoring, predicting, and preventing extreme climatological and pollution events, with the definition of new objectives of integrated development, involving human beings, the atmosphere, and nature and its wealth. A new and interesting approach should be taken, combining science, society, economy, and ethics with energy, development, and environmental protection against war, injustice, and poverty.

A new world governance is therefore needed, with stronger correlations between energy policies and pollution and climate change measures. Likewise, it should be strongly affirmed that good international relations and nonconflicts are key tools for secure energy supplies, the recovery of social and economic development, and the fight against poverty all over the world (Pontificium Consilium de Justitia et Pace, 2013).

## List of Terms

**Allochthonous:** Sediment or rock that originated at a distance from its present position.

**A.T.M.O.S.FE.R.A:** A system that allows daily forecasts of concentrations of pollutants. The software for this system is based on neural networks.

**Meteoclimatic:** A weather event linked to climate change.

# 13

## National Security and Public Health Implications of Climate Change

Climate change is an issue within a global context and should be treated as a risk to people's health. Principal concerns include injuries and fatalities related to severe weather events and heat waves; infectious diseases related to changes in vector biology, water, and food contamination; allergic symptoms, respiratory and cardiovascular disease related to worsening air pollution; and nutritional shortages related to variations in food production. Secondary concerns include mental health consequences, population dislocation, and civil conflict; however, data is scarce for these issues compared to the primary concerns.

The changing climate jeopardizes the safety and well-being of the public and requires the same treatment as other strategic threats such as cybersecurity and terrorism. They are all a danger to national security. Global health is a humanitarian endeavor that seeks to improve the well-being of the world's population, whereas national security works to protect the interests of people within a given state. To meet these challenges requires action to create decision-support systems to analyze and set priorities for how to address the health and national security implications of climate change.

The approach to climate change impacts on health shows clear similarities. We can itemize them as primary, secondary and tertiary prevention. Collectively these practices are known as public health preparedness:

- Primary prevention corresponds to mitigation—Efforts to slow, stabilize, or reverse climate change by reducing greenhouse gas emissions
- Secondary and tertiary prevention corresponds to adaptation—Efforts to anticipate and prepare for the effects of climate change, and to reduce the associated health burden

Mitigation efforts will occur mainly in sectors other than health, such as energy, transportation, and architecture (although the health sciences can contribute useful information regarding the choice of safe, healthful technologies). Adaptation efforts, on the other hand, correspond closely to conventional medical and public health practices (Frumkin et al., 2008).

It is crucial for the health sector to establish preparedness efforts for specific events, such as terrorist attacks (especially since September 11, 2001), the emergence of new infectious diseases, and the reemergence of old ones (including the possibility of pandemics such as avian influenza) and the occurrence of natural disasters such as earthquakes and hurricanes. These possibilities create a sense of urgency that should force health professionals to be ready for the events as mentioned earlier. Public health preparedness for climate change events has implications on national security and requires the same vigilant approach. As we know, events such as an influenza pandemic, a terrorist attack, or a hurricane cannot be predicted with precision. Thus, protecting public health and developing a strategic approach to minimize the effects of severe weather events remains essential (Frumkin, McMichael, and Hess, 2008).

## Potential for Conflict

The effects of climate change have the potential to increase conflict and may challenge the human security and survival. The escalation of social tension may create violent riots and domestic civil dissension, and as scarcity of resources aggravate, bilateral or regional nonviolent or violent conflicts may arise (Brauch et al. 2002).

Environmental and demographic stressors are expected to force large numbers of people to mobilize or face displacement. According to the United Nations, 22 million people were displaced in 2013 by disasters brought on by natural climate events. The Intergovernmental Panel on Climate Change (IPCC), the UN's science advisory board, estimates an increase in the number of displaced people over the course of this century (United Nations High Commissioner for Refugees, 2015). Climate change impacts contributes to the chronic and projected decline in basic resources for subsistence such as freshwater, living space, and arable land, which are factors affecting population relocation, migration, and displacement, leading to societal chaos, disease, and conflict.

Werrell and Femia (2015) explain that national security practitioners call climate change as a "threat multiplier" (CNA Corporation, 2007; U.S. Department of Defense, 2014) or an "accelerant of instability" (U.S. Department of Defense, 2010). The effects of climate change have the potential to worsen other drivers of insecurity and create "a new chain reaction of events, each caused by the previous one." In that context, climate change interacts with these other environmental, economic, social, and political factors that affect people's stability and well-being in addition to their health welfare. Climate change has already been claimed to be a factor in the Darfur clash in Sudan as drought and ecological change directly caused conflict. Any increase in conflict (and even the perception of enhanced risk) has important public health implications (Butler, 2012).

## The Intertwined Relation between Security, Health, and Climate Change: An Augmenting Force of Instability

The connection between national security, intelligence, and health has been established since the early days of World War II. The alarming threats of totalitarianism in Europe and Japan brought these problems to the forefront due to the increased compilation of relevant information about political and foreign military developments at the Federal Bureau of Investigation (FBI) and the Department of State, Department of War (dissolved in 1949), and Navy. At the strategic, operational, and tactical levels of war, the United States has previously used finished medical intelligence. All three levels were used to classify a wide range of tendencies in foreign military and civilian biomedical research and development because they could pose a threat to national security. Public health, environmental, and social factors that may possibly have an effect on military operations, disaster relief, or broader national security interests became relevant issues for consideration (Clemente, 2013).

We can start a dialogue interlinking public health, national security, and the spread of disease asking a crucial question that Professor Andrew T. Price-Smith asked in his book *The Health of Nations: Infectious Disease, Environmental Change, and Their Effects on National Security and Development*: Can infectious disease negatively affect state capacity by generating political, economic, and social instability? If so, how does infectious disease contribute to political instability and underdevelopment (Price-Smith, 2002)? We can assert that the relationship between disease and political instability is indirect but a tangible and existent threat. As the scarcity of resources accelerate, the struggle for political power to be able to control those scarce state resources intensifies, where climate change is already exacerbating the shortage of them, and should ultimately magnify the harsh economic and social impacts of infectious diseases (Noah and Fidas, 2000).

The presence of U.S. civilian and military personnel overseas is still strong and will remain at risk from infectious diseases. An outline of risks from infectious diseases and its implications for U.S. national security includes

- Emerging and reemerging infectious diseases, many of which are likely to continue to originate overseas, kill at least 170,000 Americans annually under the current climate conditions. Many more could perish during a severe influenza pandemic or yet-unknown disease (Hamburg, 2008).
- Infectious diseases account for many military hospital admissions. U.S. military personnel deployed at NATO and U.S. bases overseas are at low to moderate risk. At highest risk will be U.S. military forces deployed in support of humanitarian and peacekeeping operations in developing countries.

- The infectious disease burden will weaken the military capabilities of some countries, as well as international peacekeeping efforts.
- Infectious diseases are likely to slow socioeconomic development in the hardest-hit developing countries and regions. This will challenge democratic development and transitions and possibly contribute to humanitarian emergencies and civil conflicts.
- Infectious disease-related embargoes and restrictions on travel and immigration will cause frictions among and between developed and developing countries (Noah and Fidas, 2000).

The adverse effects of climate variability and change will significantly increase the number of deaths. The health challenges will continue to escalate and will mainly strike the urban poor in developing countries. For example, climate change will have both indirect and direct effect on many diseases

**TABLE 13.1**

Climate Change and Human Health

| Environmental Changes | Example Diseases | Pathway of Effect |
|---|---|---|
| Dams, canals, irrigation | Schistosmiasis | ▲Snail host habitat, human contact |
| | Malaria | ▲Breeding sites for mosquitoes |
| | Helmintiasies | ▼Larval contact due to moist soil |
| | River blindness | ▼Black fly breeding, ▼disease |
| Agricultural intensification | Malaria | Crop insecticides and ▲vector resistance |
| | Venezuelan hemorrhagic fever | ▲Rodent abundance, contact |
| Urbanization, urban crowding | Cholera | ▲Sanitation, hygiene, ▼water contamination |
| | Dengue | Water-collecting trash, ▲*Aedes aegypti* mosquito breeding sites |
| | Cutaneous leishmaniasis | ▲Proximity, sand fly vectors |
| Deforestation and new habitat | Malaria | ▲Breeding sites and vectors, immigration of susceptible people |
| | Oropouche | ▲Contact, breeding of vectors |
| | Visceral leishmaniasis | ▲Contact with sand fly vectors |
| Reforestation | Lyme disease | ▲Tick hosts, outdoor exposure |
| Ocean warming | Red tide | ▲Toxic algae blooms |
| Elevated precipitation | Rift Valley fever | ▲Pools for mosquito breeding |
| | Hanta virus pulmonary syndrome | ▲Rodent food, habitat abundance |

*Source:* Vandermeer, J. H., 2003, *Tropical Agroecosystems*, Boca Raton, FL: CRC Press.
*Note:* ▲ represents increase; ▼ represents decrease.

endemic to African countries such as malaria and dengue fever. As temperatures rise, the range for malaria expands and excessive flooding may indirectly lead to an increase in vector-borne diseases such as cholera through the expansion in the number and range of vector habitats. Climate change in megacities can harmfully influence all ecosystems. These climatic impacts can potentially influence water-borne and vector-borne diseases by expanding and creating conducive environments. Climate change will continue to distress all sectors of society from national and economic security to human health (Krämer et al., 2011). To what extent all these elements affect national security is related to the changes in climate patterns and their impact on the physical environment that can create profound effects on populations, especially the urban poor. These elements build new challenges to global security and stability. As health risks are interconnected to security instability, failure to anticipate and mitigate these changes increases the threat of volatility and potential for conflict in such populations with the looming danger of spreading throughout borders and nations.

Table 13.1 shows examples of how diverse environmental changes affect the occurrence of various infectious diseases in humans. Please note urbanization/urban crowding, link to megacities, climate change effects such as ocean warming and elevated precipitation and its implications are interlinked and corresponding with each other.

## Framing Climate Change as a Threat to National Security: Redefining Security–U.S. Interests at Risk

The Dutch chess grandmaster, mathematician, and author Dr. Max Euwe wrote: "Strategy requires thought, tactics require observation." Essential to understanding his words is the knowledge that any strategy needs an effective deterrence component and the ability to influence the actor's behavior. If we address a global issue such as climate change impacts, a strategy requires a multifaceted effort across all sectors of society to maximize its effectiveness and create the right environment to diminish the risks and effects. Any plan needs a thought-building process and an observational study that reach a conclusion on the basis of evidence and reasoning about the possible effects of treatment on the issue.

The correlation between climate change and public health impacts has been established by many professionals throughout many years of research. The first few chapters of this book presented a comprehensive amount of information that solidified the case for climate change and health risks to be on top of the climate talk agenda. Now, security risks add another complexity to its nexus to solve as severe weather events such as flooding, heat waves, and the spreading of diseases cause migration and consequently affect populations as a whole; this can lead to internal and international conflicts.

## Vulnerability and Risks

The events of September 11, 2001, marked a new era that unmasked the United States' vulnerability and the risks associated to external threats that affect our freedom, national and international interests, way of life, and our citizens' safety, and caused a paradigm shift in solving threats to our national security. The importance of the emerging "nontraditional" security issues that have detrimentally affected the developing world could also progressively impinge on policy decisions and national interests. These nontraditional issues such as climate change, resource scarcity, environmental deterioration, and declining productivity are now issues faced by the developed world, which is confronted with human-centered vulnerabilities that were previously nontraditional challenges for the developing world (Liotta, 2007). Due to the uncertainty of these events in the near and long-term future, a proactive approach is necessary to adapt and subsequently alleviate the impacts of climate change. As previously described, the implications of climate change in national security and health are interrelated by the risks of global instability caused as a result of food and water shortages, pandemic disease, and disputes over refugees and resources, migration, and devastation caused by severe weather events. The fragile environment, particularly in megacities in the developing world, cannot support the impacts of these harsh events, and it will lead to uncertainty, malnourishment, and social unrest, which translates into humanitarian disasters, political violence, and destabilized weak governments.

Megacities' concerns about climate change impacts will increase and might become more significant than any physical or financial damage a serious weather event might cause. The perception of rapid change will create urban insecurity, although this is not unique to megacities. It increasingly makes cities isolated and forced to take unilateral action to secure resources, territory, and other interests. Megacities are our global future with their unparalleled population and an economic boom because many are now designated free-trade zones. Shenzhen, China, was a small town until 1980 and is now a 10 million inhabitant city that ranked fourth in China for industrial output, manufacturing higher technology products with several of its own successful sunrise companies (Shenzhen Government Online, 2014). Megacities' motivation to engage in greater multilateral cooperation will depend on several factors, such as the activities of other cities or countries, the economic performance and perspective, or the importance of the interests to be defended or won. Climate change is an important dynamic to consider.

## The State of National Security

The 1947 National Security Act established the basis for American security in the Cold War. It substantially reorganized the foreign policy and military establishments of the U.S. government and, more than two decades after the fall of the Berlin Wall, the basis of its structure is still in place. British statesman Lord Palmerston described core national interests in 1848 as the "eternal" and ultimate justification for national policy (Jablonsky, 1995).

The interests of the United States can be categorized in three ways:

- Physical security
- Promotion of values
- Economic prosperity

Physical security involves the protection of the people of the nation-state against an attacker. That requires the fundamental certainty of survival ensuring the values and institutions of the nation remain unbroken (Jablonsky, 1995). Promotion of values refers to the establishment of the legitimacy of, or the expansion of, the fundamental values of the nation such as free trade, human rights, and democracy (Yarger and Barber, 1997). Economic prosperity is the combination of wealth and well-being, unpacking important "pillars" of prosperity, including economy, entrepreneurship and opportunity, governance, education, health, safety and security, personal freedom, and social capital (Legatum Institute, 2015). The three national interests are the fundamental cores of the strategic role of the United States in the world (Jablonsky, 1995). A solid conceptual foundation with a comprehensive risk management framework to make it relevant and to strategize against climate change impacts on health and its consequences on national security become imperative as the threats, once considered into a distant future, are now in progress.

## The Ultimate Security Risk

Climate change is a complex issue that is displayed at the front page of the mainstream media and draws heated arguments from different parts of society about how and when to cope with its effects. When it is coupled with national security and public health, the message delivered acquires a personal validity that can lead to an emotional attachment to the issue. Historically, climate change has been framed as an environmental problem and more recently as a political problem. Different dimensions to how the public and world leaders frame climate change needs to be altered. It is imperative to focus on security issues as the stability of global markets is now more challenged than ever. Energy sources, supply and demand, and the spread of diseases that can cross borders and nations, and possess a threat that is palpable, must be aligned to a national security strategy.

Although climate change may not create a direct national or international conflict per se, it is an accelerant of conflict and instability, adding to the effects of extreme weather events that may lead to increased demand for defense and disaster support as well as placing a burden on humanitarian assistance (Zinni, 2010).

Vital national interests are at risk, as climate change becomes a threat to national security. A direct threat is when the loss of lives or welfare of U.S. citizens is at risk due to direct environmental degradation or damage, and impairs our most important national values. As the world becomes more volatile and the threats to the United States more unpredictable, the necessity of acting urgently to be prepared in accordance to global trends, negative and positive, becomes imperative.

Security entails an assessment of risk of exposure, susceptibility to loss, and capacity to recover and a highlighted discourse on vulnerability. The nation (national security), basic needs (human security), income (financial security), and property (home security) are associated with the most important of vulnerable entities (Barnett, 2001). As history recalls, the spread of nuclear weapons and nuclear technologies have been a highlighted concern for the United States for obvious reasons, and nuclear proliferation has been an important issue of national security since World War II. Although there are many nuclear global treaties and arrangements designed to reduce the risk of nuclear war and nuclear accidents, the threat still exists. North Korea and Iran, for example, are constant threats placed on the front page of our national security agenda. Nevertheless, environmental security is also a growing concern that goes across nations with a global reach and devastating consequences (Barnett, 2001).

"The Changing Atmosphere: Implications for Global Security" was the first international meeting of scientists and national policymakers to emphasize the risks climate change poses to humanity; it was held in Toronto in 1988. The conference concluded that "humanity is conducting an unintended, uncontrolled, globally pervasive experiment whose ultimate consequences could be second only to a global nuclear war" (Barnett, 2001).

The scale of this statement should be reflected in any strategic initiative. The correlation between the effects of climate change and its implications to national security are highly relevant today and should be integrated fully into a National Infrastructure Protection Plan and Strategic National Risk Assessment with a careful analysis of the probabilities of an event and the resultant consequences of that event occurring (CNA Military Advisory Board, 2014).

## The Health Factor: Megacities' Challenges in Health Care and the Repercussions on National Security

The urban poor, who live under extremely fragile conditions and are concentrated in megacities, are exposed to perilous conditions magnified by climate change. This urban population is unstable and these particular conditions raise the threat of declining resource availability and food production. Every security problem, whether it is human security, environmental security, or national security, would be accentuated where sufficient economic development is not present (Liotta and Miskel, 2012). Megacities' urban population deficiencies and vulnerabilities magnify the detrimental effects of climate change and line up a new threat on national security with new challenges. As climate change moves toward its path and severe weather events become more intense in megacities, those dangerous elements will, in time, undermine the system in place. Scarcity of water for drinking and irrigation, outbreak and rapid spread of disease, or lack of sufficient warning systems for natural disasters or environmental impacts may be linked to security issues. Author P. H. Liotta called it "entangled vulnerability scenarios." These issues warrant a greater concern in addition to the traditional focal point on hard security "threats" (Liotta and Miskel, 2012).

## Net Zero: A Solution to Climate Change Adaptation and National Security

*Katherine Hammack*

### Climate Change and Resource Scarcity

The impacts of climate change are being felt today in cities and on Army bases across the country and around the globe. Climate change undermines and overwhelms regional ecosystems, disrupting the services they provide to Army installations and the surrounding civilian communities. The impacts of a changing climate create vulnerabilities and liabilities that increase both costs and risks to the Army and our national security.

Water availability is affected by both drought and overabundance of rainfall. Drought leads to water restrictions and impacts commercial and industrial operations, biodiversity, and air quality. An overabundance of rainfall can lead to water system failures, impact water quality, and affect biodiversity.

Climate change also affects energy supply. There is an increased risk of shutdowns at coal, natural gas and nuclear power plants. Changes in the climate mean decreased water availability, which affects cooling at thermo-electric power plants, a requirement for operation. And when water is available, higher temperatures make the cooling process less efficient. There are also higher risks to energy infrastructure located along the coasts due to sea level rise, the increasing frequency and intensity of storms, and higher storm surge and flooding.

Power lines, transformers, and electricity distribution systems face increasing risks of physical damage from the hurricanes, storms, and wildfires that are growing more frequent and intense. Energy usage increases as air-conditioning needs rise due to increasing temperatures and heat waves, along with the risks of blackouts and brownouts in regions throughout the country.

Climate change drives people to move: to evade coastal erosion and sea level rise, to follow changes in rainfall, to avoid the stress that comes with high heat. Mass migrations affect cities, which require land for expansion. The tension between land for food production and land for development is seen across the United States and more acutely in other countries around the globe. The practice of building landfills must change as land becomes even more precious. Burying resources for future generations to excavate and recycle is a practice that must end.

End of initial life plans for purchased goods should not include burial in a landfill. Elimination of the need for waste disposal in landfills is the objective. Resource recovery must be determined at the time of purchase. Cradle-to-cradle is the sustainable management of all solid resources. A product contains resources that can be recycled or reused at the end of the initial product life. Sustainable purchasing means that the total lifecycle of the product and packaging is known at the time of purchase.

Recognizing all of these factors, and the risk to national security, the Army has adopted a net zero approach that applies whole-systems design to build resiliency on our installations.

## The Army's Adaptation Strategy: Net Zero

The Net Zero strategy is the cornerstone of the Army strategy for sustainability and energy security. This strategy is based on the principles of integrated design, which will ensure the Army of tomorrow has the same access to energy, water, land, and natural resources as the Army of today. The Net Zero approach consists of five interrelated steps: reduction, repurpose, recycling and composting, energy recovery, and disposal. A Net Zero energy installation produces as much energy on-site as it uses over the course of a year. A Net Zero water installation limits the consumption of freshwater resources and returns water back to the same watershed so not to deplete the groundwater and surface water resources of that region in quantity or quality. A Net Zero waste installation reduces, reuses, and recovers waste streams, converting them to resource values with zero landfill.

The Army is working to enhance efficiencies in energy, conserve water, and reduce waste generation, as it is essential to our current security and future operational missions. The Net Zero strategy is integrated into contingency base planning, construction, and operations.

The vision is to create sustainable installations that support the missions of the Army to train as we fight and successfully protect and defend our nation's interests. The land, buildings, vehicle platforms, and infrastructure will continue to provide excellent quality-of-life support for soldiers and their families. This commitment applies to all domestic and overseas installations whether they are permanent, enduring, or temporary.

The Army launched the Net Zero initiative in October 2010, and by April 2011 had selected 17 pilot installations striving to reduce consumption of resources in one or more of three categories—energy, water, and waste—to an effective rate of zero by 2020. Net Zero is comprised of three parts (Figure 13.1):

- Net Zero energy—Reduce overall energy use, maximize efficiency, implement energy recovery and cogeneration opportunities, and then offset the remaining demand with the production of renewable energy from on-site sources, such that the Net Zero energy installation produces as much renewable energy as it uses over the course of a year.

- Net Zero water—Reduce overall water use, regardless of the source; increase efficiency of water equipment; recycle and reuse water, shifting from potable water use to nonpotable sources as much as possible; and minimize inter-basin transfers of any type of water, potable or nonpotable, such that a Net Zero water installation recharges as much water back into the aquifer as it withdraws.

- Net Zero waste—Reduce, reuse, recycle/compost, and recover solid waste streams, converting them to resource values, resulting in zero landfill disposal.

The Net Zero concept, as described, is based on a hierarchy of activities that seek reduction first and foremost—reduction through better design, improved management, reduced requirements, and behavioral or cultural change. It stresses that installations first seek low-cost and easy-to-implement activities, such as improving energy efficiency in existing facilities, implementing water efficiency practices, and eliminating generation of unnecessary waste. The concept then moves on to other activities needed to efficiently meet energy and water demand or eliminate solid waste to landfill.

Recovering energy through cogeneration or recycling water through reclaimed water systems should take precedence over the pursuit of new sources. Likewise, recycling, composting, and repurposing should take precedence over waste-to-energy conversion. Developing new energy projects and disposing of waste in landfills should be the last options. All of these

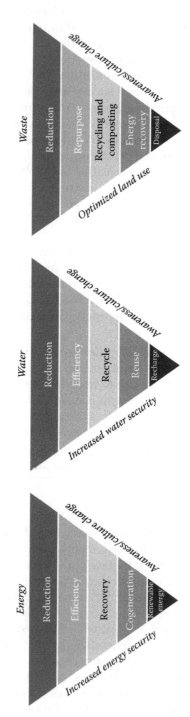

**FIGURE 13.1**
Net Zero components.

approaches must be woven into existing planning efforts and linked to ongoing programs.

### Net Zero Energy

Achieving Net Zero energy requires reducing overall energy use, maximizing energy efficiency, and implementing demand management and energy recovery practices. Once these avenues have been pursued, offset the remaining energy demand through the production of on-site renewable sources, so they produce as much energy on-site as they consume over the course of a year.

The path to Net Zero energy requires careful planning that takes into account the need for conservation awareness throughout the community. It also requires technological solutions for facility efficiency gains as well as larger-scale, on-site generation with effective management and control of those assets. Partnerships with public utilities or the private sector to support large-scale renewable energy projects, and negotiate with the private contractors that own and operate the electrical infrastructure at many installations, similar to the infrastructure in large cities.

Understanding the installation energy baseline and future growth from a master planning perspective is essential to laying out an integrated approach to Net Zero energy. As a result, Net Zero installations use cross-functional teams, including mission representatives, privatized utility providers, and residential community managers, to provide input for planning as well as current construction. Roadmaps are developed, through a series of charrettes, which articulate the ultimate objectives, timelines, and specific steps to achieve the desired end state.

Education and awareness are important ongoing activities to achieve Net Zero energy. Engagement within the community involves existing staff, private utilities, and building occupants. Facility managers can appoint building energy monitors (BEMs). BEMs leverage the institutional knowledge of the building occupants to identify and eliminate energy waste. This approach also can enable competitions between like buildings for the greatest progress. BEM programs involve the building occupant in the energy and facility upgrade. They have input into standard operating procedures in the building. Behavior change, when initiated and supported by building occupants, can result in long-term efficiencies. The Department of Energy estimates that behavior change can reduce energy costs by 3% to 10% per year. For the Army, that would mean annual savings of $40 million to $100 million.

Net Zero installations have also leveraged metering data to develop energy consumption reports for end users. These reports compare current usage to the past year as well as industry standards. The American Society of Heating, Refrigeration and Air-Conditioning Engineers (ASHRAE) in conjunction with the Department of Energy has developed benchmark Energy Use Intensities (EUIs) by building type and climate zone. ASHRAE Standard

100 charts EUIs in annual energy use per square foot (kBtu/ft$^2$/yr) for 53 types of buildings in 17 different climate zones. These industry standards can identify target end states for buildings of any age and type.

In residential neighborhoods on a military base, soldiers do not directly pay their energy bills. Billing programs have been developed to compare energy usage to that of neighbors in similar housing. If your energy usage is outside of the average range for similar houses, you get a bill for the differential. If your energy usage is significantly under that of your neighbors, you get a credit. Outreach and education programs in residential communities seek to include service members' families in an installation's Net Zero efforts. Results are a 15% to 23% reduction in energy consumption.

Pilot sites have relied heavily upon facility upgrades and new technologies to meet their goals. For efficiency projects and on-site generation development, partnerships with industry and utility providers are a critical component of success. The Net Zero sites have leveraged multiple federal authorities that take advantage of public–private partnerships and third-party financing options to invest in upgrading energy infrastructure and increasing onsite generation. Some examples include energy savings performance contracts (ESPCs), utility energy savings contracts, and enhanced use leases (EULs) or power purchase agreements (PPAs). Installations have used these partnerships to fund streetlight replacements, lighting retrofits, occupancy sensors, energy management control systems, substation and line upgrades, microgrids, water leak repair, building envelope upgrades, boiler upgrades, and window replacements.

Renewable power generation is the last stage in the Net Zero journey. PPAs are long-term contracts for renewable energy. The Army does not design, own, maintain, or operate these systems. Power is purchased from a third party at or below local retail rates for electricity. A commitment was made to the president of the United States to generate 1 GW of renewable energy on Army bases by 2025. The Army is choosing to meet this commitment by ensuring all such projects are focused on generating power that is connected directly to the installation's power distribution grid, increasing the reliability and resiliency of local power networks. In some cases, these projects come with sophisticated power management features that allow Army installations to operate independently from the wider, regional grid.

The primary intent of Net Zero energy is to ensure the military mission can be sustained over the long term. In times of crises or natural disasters, installations become emergency management centers, providing support to surrounding communities. Net Zero is mission critical.

### Net Zero Water

Achieving Net Zero water requires an installation to reduce overall water use, regardless of the source. It also requires increased reliance on technologies that use water more efficiently; shifting potable water consumption to non-potable sources as much as possible; and minimizing interbasin transfers of any type of water, potable or nonpotable, so that the installation recharges as much water back into the local aquifer as it withdraws.

Water balance assessments and project roadmaps can guide progress (Figure 13.2). A water balance assessment identifies current water supply and estimated water use for each major end-use including potable and industrial, landscaping, and agricultural water applications. The project roadmaps used the water balance results to create a strategy to meet the reduction goals and site-specific Net Zero water objectives. Each roadmap contains water conservation measures (WCMs) and a life-cycle cost analysis for each WCM to determine the economic effectiveness of each measure, both individually and as part of bundled projects.

Successfully implementing these strategies requires behavioral change and conservation awareness throughout the installation community. The low cost of water makes for a longer return on investments. But cost may not be the sole driver of Net Zero water.

Many of the water systems that keep ecosystems alive and feed a growing human population have become stressed. Rivers, lakes, and aquifers are drying up or becoming too polluted to use. More than half the world's wetlands have disappeared. Agriculture consumes more water than any other source and wastes much of that through inefficiencies. Climate change is altering patterns of weather and water around the world, causing shortages and droughts in some areas and floods in others. At the current consumption rate, this situation will only get worse. By 2025, two-thirds of the world's

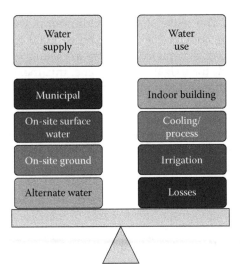

**FIGURE 13.2**
Water balance assessment.

population may face water shortages. And ecosystems around the world will suffer even more.

As installations strive to reach Net Zero water goals, educating the base population is vital. Net Zero pilots have created new water conservation policies, councils for water stakeholders to share information and identify priorities, and award programs to recognize exemplary efforts by individuals. Awareness can be focused externally, such as participating in state-level water conferences to learn about regional water concerns, or internally, such as publishing articles in the post newspaper about Net Zero water efforts. It also requires effective and ongoing partnerships with utilities—both municipal and privatized utility providers on the installation.

Many installations have integrated Net Zero water seamlessly into their existing sustainability efforts. Maintenance departments focus on ensuring that the right kind of water sense fixtures are obtained and operate properly. Should a water shortage occur, plans are made to minimize the impact of a water supply shortage on the base operations, providing first for public health, fire protection, and other essential services.

ESPCs are becoming more involved in the financing of water conservation and efficiency projects. Installations have leveraged these funding mechanisms to overcome shortages in operation budgets. The ESPC contractor becomes the installation's partner in plumbing retrofits. These upgrades—low-flow toilets, low-flow showerheads, and faucet aerators—are low-cost improvements that, when bundled together, can produce significant water and energy cost savings over time, drawing the interest of ESPC contractors.

Alternative water sources have proven essential for the Net Zero pilots to reach their goals. The reuse of water within wastewater treatment plant operations is saving Tobyhanna Army Depot 5,000 to 6,000 gallons per month in potable water. Several installations are using water reclaimed after wastewater treatment for irrigation. For some, this means expanding existing systems with more pumps, pipes, and storage ponds to increase their capacity to use reclaimed water.

Municipalities in water-scarce regions may have reclaimed water that the installations could procure to help reduce potable water demand. An installation's Net Zero objectives may provide a utility the needed incentive to prompt greater investments in water reclamation infrastructure outside the fence.

Similarly, installations that have privatized utilities will rely on partners to achieve success. Replacing water fixtures to increase efficiency is a large part of Net Zero, but other strategies need to be explored as well. Leak detection and repair also play a role and, for utilities that are privatized, this must be done through existing contracts. Management approaches to ensure adequate pressure and flow also can play a role, again with the help of the private utility partner.

To date, the Army has demonstrated significant progress in reducing water consumption. Overall the Army has reduced its potable water consumption by 29% since 2007. In drought-impacted California, the Army has reduced its water consumption at Sierra Army Depot by 64%, Fort Irwin by 53.5%, and the Presidio of Monterey by 21%.

Coordination is essential to ensure changes made by the installation, its tenants, and the utility all work in concert and do not result in unintended outcomes. For instance, fire suppression, backup power generation, and other mission-critical functions that depend on water must be factored into utility planning, capital improvements, and Net Zero efforts. Integrated Net Zero—pursuing the initiative in energy, water and waste—is critical. It takes energy to pump, treat, distribute, reclaim, process, and dispose of water.

### Net Zero Waste

Achieving Net Zero waste requires an installation to reduce, reuse, recycle, compost, and recover solid waste streams, converting them to resource values that result in zero landfill disposal. The Army's successful Net Zero waste initiatives have involved local municipalities, businesses, nonprofits, and the community as a whole.

Net Zero waste programs will require a focused effort to successful recycling, reuse, and composting. There are differences by region, dependent upon the local community's waste stream. When efforts and priorities are combined, new markets and other opportunities emerge. An installation's contributions to new diversion or recycling efforts potentially can provide the tipping point to turn these waste streams into viable markets for the municipality or local entrepreneurs.

Joint Base Lewis-McChord (JBLM) in the state of Washington, is turning organic waste resources into valuable landscaping and soil amendment products. Since 2006, JBLM has been diverting organic wastes such as leaves, grass, food waste, landscaping and land clearing debris, storm debris, stable waste, and biosolids (sewage sludge) from an off-base landfill disposal and composting it. During this period, JBLM has diverted approximately 28,905 tons of organic wastes. A conservative estimate of cost savings for the amount of diverted organic waste is approximately $2.6M.

Reviewing the range of wastes commonly found in municipal wastes indicates that most can be repurposed, reused, recycled, or composted in some form or another. Given the diversity of waste and options for reducing, repurposing, and recycling waste, selecting the right mix of solutions, first requires an understanding of inputs coming on the installation, the conversion of these products into waste, and the management of that waste.

Material flow analysis (MFA) is a systems-based perspective for understanding the flows and stocks of materials within a system. With the scale of an installation, the complexity and the diversity of material and waste streams, and the lack of readily available data, a complete MFA for an installation is not practical, nor is it necessary for achieving Net Zero waste. However, the principles of MFA can be employed to engage stakeholders; assess how inputs, activities, and outputs contribute to waste; identify specific situations where MFA is warranted; and, ultimately, inform strategies and roadmaps for achieving Net Zero waste.

Many different organizations on an installation procure material inputs, perform activities that convert some portion of these material inputs into waste, and distribute the resulting waste to be managed. Understanding where and what type of waste is generated within this complex system helps decision makers develop an informed strategy and select the right mix of solutions for achieving Net Zero waste. Material flow information (including solid waste data) can be organized by installation activities. For each activity group, such as dining, utilities, administration, or research, a set of organizations responsible for generating most of that activity group's waste can be identified. Throughout the survey, material flow data should be organized both by activity group (at the organization level) and by material group (at the material category level).

The purpose of a material flow survey is to collect information about materials used across the installation, the resulting waste, methods used to manage the waste, and the extent to which waste can be diverted. The types of information collected during a material flow survey include

- Information about material inputs, procurement, and supply data
- Waste characterization study of waste streams going to landfill (primary)
- Information on solid waste diversion and management from a survey of many organizations
- Specific data sets from existing key solid waste and diversion processes

The qualitative information from the material survey provides a detailed picture of the waste story at the installation. A divertability scale can be developed to classify the qualitative information about material flows and help tell the installation's waste story (Figure 13.3). The relative ease by which each material category can be diverted on an organization-by-organization basis should be assessed using information collected during the material flow survey as well as general information about existing waste diversion technologies, infrastructures, and markets.

Another option to evaluate is waste to energy (WTE). The viability of this option depends upon the volume of the solid waste stream. Most commercially viable WTE systems require large quantities of waste (>300 tons per day). When an installation is maximizing recycling, this reduces the amount of waste while removing many combustibles from the waste stream. The viability of a WTE is dependent upon the volume of waste, Btu content of

| Score | Status | Description |
|-------|--------|-------------|
| 1 | Diverted | Diversion program is in place and estimate at least 90% compliance. |
| 2 | Underutilized program | Diversion program is in place, but less than 90% compliance. |
| 3 | Opportunity | Diversion possible through adoption of existing technologies, infrastructure, markets, etc. |
| 4 | Challenge | Diversion possible, but will require additional development of available technologies, infrastructure, and/or markets. |
| 5 | Problem | Diversion would require development of new technologies, infrastructure, and/or markets. |
| 0 | Incomplete information | Incomplete information about presence and/or destination of waste stream. |
| | No waste | Waste stream does not exist for this organization. |

**FIGURE 13.3**
Telling the story with a divertability scale.

the waste, permitting limitations, the cost of energy, and the avoided cost of tipping fees.

Addressing specific waste streams, such as food waste, can have beneficial results. Nonprofit agencies may accept food donations, reducing a source of organic waste. The municipality may have a composting program that could accept discarded organic materials. Federal partners are also a critical part of the Net Zero waste initiative. The Army and Air Force Exchange Service (AAFES) has been an active partner in waste reduction by changing shipping and packaging disposal techniques and donating to food banks. Something as simple as eliminating plastic shopping bags starts with AAFES support and involvement within the community.

## Conclusion

Achieving Net Zero requires attention on the ultimate objective: resource management in a sustainable manner. When the entire community of stakeholders associated with an installation is involved, Net Zero gets closer to reality. Implementing Net Zero practices cannot be viewed solely as the responsibility of facility managers. There are many challenges the community and private sector can address best when working together.

Net Zero energy, water, or waste are difficult to achieve on an individual building basis. When considered at a campus or city level, the Army experience has demonstrated that it becomes more achievable. Buildings come in all different shapes and sizes, from shacks to off-grid hotels to traditional style homes, to commercial buildings that involve more complex systems. Some have very high energy and water loads; some are high waste generators while others can be net zero "plus." A true Net Zero base is becoming more attainable financially with the continued advancement and affordability of building technology.

Loss or restriction on the use of training lands attributed to climate factors incurs real costs in terms of time, money, and resources. Without predictable access to training areas and ranges, individual skills and unit readiness will suffer. In turn, the Army's ability to respond when called upon to meet the needs of the nation is affected, making us more vulnerable at a time when we can least afford it.

Energy supply shortfalls, coupled with water scarcity, represent a strategic vulnerability for the Army—increasing the risk to the mission, pocketbook, and reputation. Ensuring energy and water security, through increased efficiencies and careful management of resources, reduces this vulnerability and increases national security.

The Army is doing its part to mitigate and adapt to the impacts of climate change. Through the Net Zero initiative, the Army has made significant progress toward the sustainability and resiliency of its installations ensuring they can continue operations, deploy soldiers, and support their local communities in case of a natural disaster.

For the U.S. Army, the existence of the impacts caused by climate change is not a theoretical debate. It is a reality to which we must adapt. The Army's Net Zero approach ensures that installations are ready and resilient over time, able to support the Army's mission. The security of our soldiers—and our nation—depends on it.

# Section IV

# Case Study: Researched Megacities

# 14

## Overview of Megacities Used in Case Study

Megacities are important centers of population concentration and growth, and derive their prosperity and importance through their long-standing commercial connections with the rest of the world. Most of these cities are located on or near a coast, which facilitates the flow of trade and contributes to their prosperity. Others are located near major rivers that serve as a channel for commerce (De Sherbinin et al., 2007).

Megacities are confronting many issues related to a changing climate and, as highlighted in earlier chapters, the vulnerabilities are much greater among coastal cities. Sao Paulo and Rio de Janeiro are good examples. The two are located on Brazil's long coastline (8,000 km or 4,900 miles). Climate change exposes these cities to environmental risks, such as sea-level rise, severe storms, floods, landslides, vector-borne diseases that are a consequence of rising temperatures, variability in rainfall patterns, and severe weather events (Costa Ferreira et al., 2011).

This research focused on the cities of New York City, Rio de Janeiro, Beijing, Los Angeles, London, Mumbai, and Lagos to strategize adaptation and mitigation plans. These seven megacities are considered highly vulnerable to the impacts of climate change in ways that will affect the urban population's health. New York and Rio de Janeiro have similarities because they are located in coastal zones, have dense urban populations, dynamic economies, and global trade resources. Therefore, these cities, and others located in coastal regions, should be a priority for social and environmental policies that seek forms of mitigation and adaptation (Costa Ferreira et al., 2011).

Los Angeles has a vast territory with a dense population that is highly vulnerable to heat waves, windstorms, and flooding. The county of Los Angeles encompasses a diverse ecosystem that makes adaptation and mitigation to climate change a priority to minimize threats to such a large and diverse population (Kersten et al., 2012).

Beijing is one of the 20 most populated metropolises in the world—and one of the most heavily polluted. This situation presents a challenge to its leaders as greenhouse gas (GHG) emissions keep rising, and millions of residents are exposed to health risks and subsequently a greater morbidity and mortality rate (Zhao, 2011).

Accentuating the differences among these seven major cities are income distribution (London, New York, and Los Angeles have higher-income populations), crime rates, and radical social inequalities (particularly in Rio de Janeiro, Lagos, and Mumbai). Other important issues are poverty and environment traps, which keep the underprivileged locked in unsuitable housing, and the increase of land fragmentation in densely populated areas. The effects of a changing climate on urban populations can also present inequities regarding preparedness, adaptation, mitigation strategies, and resources to combat climate change.

Health is closely related to a population's socioeconomic status. For example, Rio de Janeiro, and Brazil in general, have a large number of slums, and consequently, underprivileged urban residents are disproportionately affected because the bottom of the social scale has poorer health than at the top (Szwarcwald et al., 1999). Thus, the results of this case study research can be presented as a model for other megacities to adopt, and to develop and implement strategies dealing with climate change and its consequences for the urban poor's health.

---

### List of Terms

**Environmental (or ecological) trap:** A situation in which rapid environmental change forces humans (and other animals) to live or remain in poor-quality habitats.

**Land fragmentation:** A phenomenon in spatial planning of cities. It is a significant factor in urban sprawl. Development can lead to disconnected communities and to new, conflicting land uses.

**Poverty trap:** A circumstance that causes poverty to persist; thus, the poor remain poor. Lack of capital and credit are at the root of the problem.

# 15

## City of New York

Over the past three decades, the global mean temperature has increased due to natural and anthropogenic factors. In the United States, the average temperature has risen by over 2°F since 1970. At the same time, precipitation has increased an average of 5%, sea level has risen, and drastic weather events (e.g., hurricanes) happen more frequently (Karl et al., 2009). As the effects of climate change continue to impact urban areas, the projections for the city of New York are an increase in average temperature of 1.5°F to 3°F by the 2020s; 3°F to 5.5°F by the 2050s; and 4°F to 9°F by 2080. These temperatures will vary accordingly to the greenhouse gas (GHG) emissions level (DeGaetano and Tryhorn, 2011) and will directly impact New Yorkers and their health.

## Rising Sea Level

Thermal expansion of the oceans will cause rising sea levels. The melting and discharge of ice from mountain glaciers, ice caps, and much larger ice sheets located in the Antarctic and Greenland will contribute significantly to this rise and are a result of anthropogenic climate effects (Schellnhuber et al., 2012).

Thus, rising sea levels are extremely likely. Global Climate Model (GCM) projections for mean annual sea level rise in New York City are 6 to 13 centimeters by the 2020s; 18 to 30 centimeters by the 2050s; and 30 to 58 centimeters approximately by the 2080s. The GCM, however, does not monitor all contributing factors to sea-level rise. The "rapid ice-melt" approach implies sea levels possibly may rise between 104 and 140 centimeters by the 2080s (Rosenzweig et al., 2009).

Figure 15.1 shows a black line representing a combined observed and projected temperature, precipitation, and sea-level rise. The three lines in the Central Range area (A1B, A2, B1) represent the average for each emissions scenario across the 16 GCMs (7 in the case of sea level). The central range is represented by the shaded area, and each year's minimum and maximum projections throughout the set of simulations are shown by the bottom and top lines on the graph. A 10-year filter has been applied to the observed data and model output. The spotted area for the years between 2003 and 2015 stands for the phase that is not covered due to the smoothing procedure (2002–2015 for sea-level rise) (Rosenzweig et al., 2009).

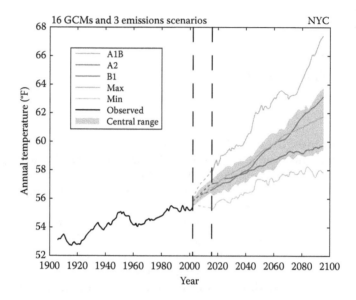

**FIGURE 15.1**

Observed climate and future projections for New York City. (From Rosenzweig, C. et al., 2009, *Climate risk information*, New York City Panel on Climate Change.)

## Ozone-Related Health Impacts

Climate change influences public health because it produces high concentrations of ambient air pollutants. For example, a study of 31 counties in New York State about the effects of elevated ozone levels on public health projected a median increase of 4.5% in summer mortality rates by the 2050s as compared to the 1990s (Bell et al., 2007).

## Urban Heat Effect

An urban heat island (UHI) magnifies the impact of warmer temperatures and affects a population's health. New York City is particularly vulnerable to UHI.

Urban heat islands form when surfaces of concrete, asphalt, metal, and stone absorb incident sunlight during the day and reradiate it as heat, which is felt most strongly at night (Knowlton et al., 2007). The result is areas of higher surface and near-surface air temperatures compared with suburban and rural areas (Knowlton et al., 2007). Many factors intensify the vulnerability of different demographic groups to heat waves and severe weather events. New York City is a diverse metropolitan area with an adult population of millions of residents, including many with cardiovascular or respiratory illness, increasing their susceptibility to summer heat stress (Knowlton et al., 2007).

The urban heat island effect is a complex meteorological phenomenon that is influenced by several conditions, including urban land use, which may affect cloud cover and wind speed. Land use and local climate also play roles in precipitation patterns, and those conditions model the impact of the UHI effect and impact on the city (Rosenthal et al., 2007). New York City has experienced the UHI effect since at least the end of the 19th century. In the 20th century the city experienced climate variability that impacted the health of its habitants. The cause of the variability was the UHI effect and the drift resulting from global warming of the earth's lower atmosphere. Over the past century the temperature has risen approximately 2°F (1.1°C), according to the Columbia Earth Institute's Metropolitan East Coast Study (Rosenthal et al., 2007).

## Sudden and Disruptive Climate Change

When Svante Arrhenius, winner of the Nobel Prize in Chemistry 1903, made the first comprehensive quantitative estimate of how human-caused emissions of carbon dioxide ($CO_2$) would modify and influence the climate, many scientists at the time thought the change would be slow. Therefore, society as a whole would have plenty of time to respond to the changes by developing and applying deterrent policies (MacCracken et al., 2008). Arrhenius published his results in 1896 (between 10,000 and 100,000 calculations). The Nobel Prize winner concluded that reducing $CO_2$ levels would decrease earth's temperature by 4°C to 5°C. In contrast, doubling $CO_2$ levels would trigger a rise of about 5°C to 6°C (Arrhenius and Waltz, 1990). He was a pioneer in the study of climate change and its environmental effects. He preceded the UN-sponsored Intergovernmental Panel on Climate Change (IPCC) comprehensive assessment by 100 years. The IPCC projected that ongoing dependence on fossil fuels for most of the world's energy will generate an increase in the global average surface temperature of approximately 1.4°C to 5.5°C (about 2.5°F to 10°F) by the year 2100. Add this change to the observed increase of about 0.6°C (about 1.1°F) over the course of the 20th century (MacCracken et al., 2008).

## The Destructive Path of Hurricanes

Millions of residents living in megacities have already experienced the catastrophic effects of climate change, leading researchers to the following question: Are there trends in extreme weather-event destruction? Hurricanes with destructive potential have increased in frequency, according to an identified trend in an accumulated annual index of power dissipation in the North Atlantic and western North Pacific since the 1970s (Emanuel, 2005).

A look at the history of hurricanes shows their intensity and destructive force. Although hurricanes are neither the largest nor the most violent storms on the planet, they combine their size and strength to become very powerful storms. These spiral masses need a complex combination of atmospheric processes to grow, mature, and then die. They are among the most lethal and costly natural disasters, with vast implications for humanity. The deadly Galveston, Texas, hurricane of 1900 was one of the most destructive storms the United States has experienced, as was Hurricane Mitch, which killed more than 11,000 people in Central America in 1998. Developing countries experience much deadlier outcomes than the United States due to the lack of preparedness, adaptability, infrastructure, and regulations. These make landfall much less frequently in developed countries, but richer countries are experiencing new and escalating levels of property loss because of booms in construction and urbanization in hurricane-prone areas (Emanuel et al., 2006). The death toll from Hurricane Katrina was 1,833 people, and this storm brought other consequences related to social and economic factors that will prevail for decades (King and Anderson-Berry, 2010). The financial losses were astonishing, and they were estimated as follows: $21 billion on structural damages, $36 billion on commercial equipment, $75 billon on residential structures and content damages, $231 million on electric utility damages, $3 billion on highway damages, $1.2 billion on sewer system damages, and $4.6 billion on commercial revenue losses. The consequences of the loss of life and damage to water systems and the environment were not estimated, but they amount to a large economic loss as well. The total economic cost estimated by risk management was at $125 billion and the estimations from Washington politicians reached a total cost of approximately $200 billion (McGee, 2008).

Figure 15.2 shows projections for rainfall and frequency of extreme storms. The figure shows an increase of the amount of rainfall. The number of years between such storms, which is called the return period, shows a projected decline. Consequently, the severity of the storms will be more intense and more frequent. These conclusions, from the UK Met Office Hadley Centre Climate Model Version 3 (HadCM3), are in accordance with previous observations of 15 other GCMs used by the Integrated Assessment for Effective Climate Change Adaptation Strategies in New York State (ClimAID).

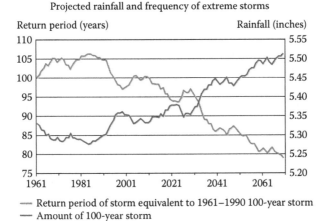

Projected rainfall and frequency of extreme storms

— Return period of storm equivalent to 1961–1990 100-year storm
— Amount of 100-year storm

**FIGURE 15.2**
The Projection for the "100-Year Storm" (1961–2061). (From DeGaetano, A., and Tryhorn, L., 2011, Responding to climate change in New York State: The ClimAID Integrated Assessment for Effective Climate Change Adaptation in New York State. *Annals of the New York Academy of Sciences*, 1244(1).)

## Megacities and Megacatastrophes

New York City and the adjacent regions of northern New Jersey and Long Island, New York, are located at a vulnerable point and exposed to hurricane storm surge. This region is particularly influenced by tides and weather because of its geographic composition of narrow rivers, estuaries, islands, and waterways. The metropolitan region is less than 5 meters above mean sea level (MSL). The risk from storm surge flooding has a range of 260 square kilometers, and the sea walls that surround lower Manhattan are only ~1.5 meters above MSL, making the area highly susceptible to severe weather events. For example, if a Category 3 hurricane hit New York City, it is projected that nearly 30% of the south side of Manhattan would be flooded, with catastrophic consequences for the infrastructure of the city and the well-being of the population (Colle et al., 2008).

Accounts show the region has been struck by extreme storms in the past, such as the New York City 1921 hurricane, which produced a 4 meter (13 foot) wall of water flooding lower Manhattan as far north as Canal Street. In 1938, Long Island experienced the impacts of the "Long Island Express" storm, with flood heights of 3 to 3.5 meters (10 to 12 feet) and up to 5.2 meters (17 feet) in southern New England. The estimated death toll was approximately 700 people. Also, Hurricane Donna, with a 2.55 meter (8.36 foot) water elevation, flooded lower Manhattan to West and Cortland Streets in 1960.

These historical events and the statistics presented show that New York City is in a quite vulnerable situation, particularly because much of the seawall that protects lower Manhattan is only about 1.5 meters above mean sea level. Therefore, a comprehensive risk management strategy addressing each of these hazards and their combination needs to be urgently developed.

Risk mitigation frameworks and related decision making will provide a path to minimize hurricane impacts on communications, transportation, roads, rail networks, and people. However, in places like New York, which is threatened by high-intensity, low-frequency hurricanes, risk assessments based on traditional data may be irrelevant to accurate projections because historical records of hurricane storm surge at a particular location are not sufficient to make meaningful estimates of surge risks (Lin et al., 2010).

---

## The New Normal: Hurricane Sandy Battered New York City

The increased frequency of severe weather events in the United States, such as the recent Hurricane Sandy, which battered New York and New Jersey, raises questions about how climate change is linked to the trend of these monster storms. Kevin Trenberth, senior scientist at the National Center for Atmospheric Research in Boulder, Colorado, stated: "The answer to the oft-asked question of whether an (extreme) event is caused by climate change is that it is the wrong question. All weather events are affected by climate change, because the environment in which they occur is warmer and moister than it used to be" (Betts, 2012).

Hurricane Sandy landed near Atlantic City, New Jersey, on October 29, 2012. This destructive storm caused 113 deaths, left up to 40,000 people homeless, and the economic losses are estimated at $50 billion, with approximately 200,000 homes damaged (Neria and Shultz, 2012).

The storm changed form from a tropical to a posttropical cyclone. It had wind speeds that reached 128 kilometers (79 miles) per hour and a central minimum pressure that reached 946 millibars at landfall. In addition, Hurricane Sandy generated tropical storm force winds extending nearly 805 kilometers (500 miles) from its center, intense flooding, destructive wind power, and power outages that immobilized and affected millions of residents, which made it a catastrophic weather event without precedent in the New York area (National Oceanic and Atmospheric Administration, 2012). According to the IPCC report on coastal systems and low-lying areas, throughout the 20th century the global rise in sea level and warmer temperatures contributed to more frequent coastal inundations, erosion and ecosystem sea ice loss, melting of permafrost (soil or rock that ordinarily remains permanently frozen), associated coastal retreat, and continuing coral bleaching and mortality (IPCC, 2007). Globally, 120 million people

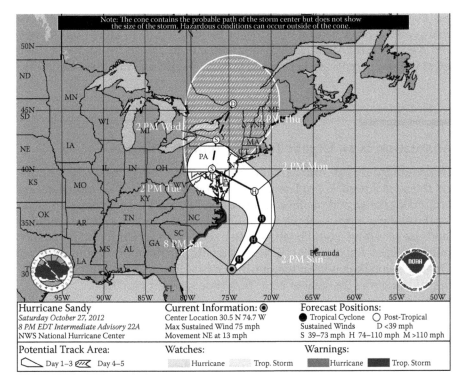

**FIGURE 15.3**
Path of Hurricane Sandy. (From National Hurricane Center, 2012, Sandy graphics archive, National Oceanic and Atmospheric Administration.)

are exposed to tropical cyclone hazards every year, which killed 250,000 people from 1980 to 2000 (Nicholls et al., 2007). Figure 15.3 shows the path of Hurricane Sandy as it pushed up the East Coast of the United States (National Hurricane Center, 2012).

## The Tragic Similarities between Katrina and Sandy

The similarities between the two devastating Hurricanes Katrina and Sandy are noteworthy and of urgent concern for climatologists, scientists, city leaders, and public health officials alike. Millions of lives were affected by flooding and its harmful effect on dense coastal populations, financial losses were in the billions of dollars; deaths, injuries, and health issues resulted; as well as the loss of electric power paralyzed businesses and infrastructures (Voiland, 2012). The changing composition of the atmosphere, including its greenhouse gas and aerosol content, is a major internal forcing mechanism of climate change. Because of natural disasters of the magnitude of Hurricanes Sandy and Katrina, it is becoming ever more difficult to defend the idea that these severe climate events are not caused by climate change (Van Ypersele, 2013).

## Adaptation Is an Urgent Topic in Today's Climate

Hurricane Sandy's destructive power taught local and global coastal dwellers the urgent need to implement an adaptation strategy. The incorporation of climate change and sea-level information into land-use planning is crucial for developing a practical plan for cities like New York. A good example of land-use planning would be to establish setback zones a minimum distance from the shore for new coastal developments. Furthermore, optimal timing for adaptation depends on monitoring coastal hazard zones and coordinating efforts across the state of New York (Buonaiuto et al., 2011).

## New York Adaptation and Mitigation Programs

As the effects of a changing climate on human health become easier to understand through accurate data and scientific reports, the need for adaptation and mitigation and a way to operate in the evolving environment is vital for survival. Former New York City Mayor Michael Bloomberg, with funding from the Rockefeller Foundation, convened the New York City Panel on Climate Change (NPCC) in August 2008. The NPCC advises the mayor and the New York City Climate Change Adaptation Task Force on matters related to the risk to infrastructure posed by climate change (Horton et al., 2009).

## New York City Government: Leading Example

New York City's ambitious plan to reduce municipal GHG emissions 30% by the year 2017 is the result of a long-term vision for a framework that affects several sectors of the government. The implementation of the framework has different priorities highlighted as follows:

> Building efficiency—The city is implementing a comprehensive strategy to reduce GHG emissions for existing buildings with a goal of at least 57%. The initiative includes energy audits, operation and

maintenance improvements, retrofits, and ongoing data analysis (Maron, 2012).

Energy audits and retrofits—The average age of New York City buildings is 60 years old, and this plan presents an opportunity to replace inefficient lighting; and cooling, heating, and ventilation systems (Maron, 2012). Identifying and implementing energy conservation measures for the city's buildings over 4,600 square meters (50,000 square feet) is a priority that can effectively improve the existing energy systems and minimize GHG emissions. "An energy audit is an inspection, survey and analysis of energy flows for energy conservation in a building, a process or system to reduce the amount of energy input into the system without negatively affecting the output" (Consensus Energy, 2012). The city's audits follow ASHRAE Level II energy audit guidelines, and it covers electrical, ventilation, and building envelope systems to determine energy retrofits and clean energy installations (Maron, 2012).

Operations and maintenance—Monitoring energy use, maintaining equipment, and operating a building's ventilation, lighting, heating and air-conditioning is an efficient way to decrease GHG emissions and reach the city's goal of 12% reduction by 2017 (Maron, 2012). The city of New York is implementing an energy efficiency operations and maintenance plan to reduce its GHG emissions impact that includes three main objectives:

- Efficiently repair, maintain, and operate existing equipment
- Improve skills and raise awareness through training and outreach programs
- Accountability and transparency (Maron, 2012)

Clean distributed generation—Clean distributed generation (clean DG) technologies require on-site energy generation that uses clean or renewable fuel sources to produce electricity and, in some cases, steam or hot water. Clean DG uses a variety of renewable energy sources such as solar thermal, wind, biomass technologies, and solar photovoltaic (Maron, 2012).

Cogeneration or combined heat and power (CHP)—Thermal and electric energy is combined to improve a building's energy efficiency and reduce GHG emissions. This technology improves grid constraints, creates a cost-saving energy program, and provides operational independence. Nevertheless, the city is implementing a string of programs that include cost-effective renewable energy applications and new technologies to exploit solar, wind, and biomass sources (Maron, 2012).

Wastewater treatment plants—The potential to reduce GHG emissions using wastewater treatment plant projects is promising. The negative aspect of this plan is that it releases a significant amount of methane gas (one of the strongest GHG emissions sources). However, the city is making general efficiency improvements and fixing methane gas leaks by recapturing methane to power electricity-generating equipment (Maron, 2012).

Vehicle fleet—The city is greening its fleet by accelerating the purchase of more fuel-efficient vehicles, mandating the use of biodiesel, adopting best practices to economize vehicle-kilometers traveled, improving vehicular deployment and fleet reduction programs (Maron, 2012).

Street lighting—Efficiency in energy reductions through the installation of 250,000 lower-watt streetlights fixtures. The decrease of energy consumption has reached 25% since fiscal year 2006. The investment was $65 million, and with reductions that have led to the total elimination of more than 40,000 $MgCO_2e$ since 2007, a full return on investment (ROI) is expected in approximately 5 years (Maron, 2012).

$CO_2$ reduction—30% annual reduction from 2006 base year by 2017 (Maron, 2012).

---

## Developing Climate-Resilient Water and Wastewater Infrastructure in New York City

*Alan Cohn*

New York City's reservoirs, some located more than 200 kilometers (nearly 125 miles) from the city, provide over 1 billion gallons of clean drinking water per day to the taps of 9 million New Yorkers. The city has consistently demonstrated superior drinking water quality—carefully managed through a Watershed Protection Program for the 5,180 square kilometer of land (approximately 2,000 square miles) of land surrounding its reservoirs—prompting a waiver from the U.S. Environmental Protection Agency's filtration requirements. The waiver, granted only to four other large U.S. cities, avoids a multibillion-dollar investment in treatment and further economic and environmental impacts associated with operations of a filtration facility. The carefully managed system taps into the built-in flexibility of its 22 interconnected reservoirs and controlled lakes. Maintaining this balance requires optimization of the system with investments in maintenance and improvements that consider an aging infrastructure, population growth, public health regulations, and fluctuations in climate.

From the tap to the drain, wastewater is collected and conveyed by 12,000 kilometers (7,456 miles) of sewers and 96 pumping stations to 14 wastewater treatment plants—all located at low points on the waterfront to minimize the need for pumping—before being discharged after treatment directly to New York City waterways. Rainfall that is collected from approximately 60% of New York City's land area is conveyed along with wastewater to treatment plants, while over the remaining land area runoff is conveyed directly to waterways. The city's treatment plants are capable of handling twice the volume of wastewater on a wet day as they do on a dry day; however, to protect biological treatment processes and avoid damage to wastewater facilities, once volume exceeds the system's capacity, the mix of wastewater and runoff is diverted directly into waterways. New York City is working to reduce these events, known as combined sewer overflows, by optimizing the sewer system and investing in green infrastructure, a decentralized network of porous pavement, green roofs, bioswales, and other technologies that avoid or slow the infiltration of runoff into the sewer system.

Recognizing the sensitivity of these systems to climate change, in 2003 the New York City Department of Environmental Protection (DEP)—the agency responsible for providing drinking water, drainage, and wastewater services for the city—began working with the Columbia University Center for Climate Systems Research, the Earth Institute, and the NASA Goddard Institute of Space Studies at Columbia University. The purpose was to develop customized regional climate change projections for DEP's planning and engineering efforts. The following year, DEP convened a formal Climate Change Task Force to oversee the department's investigation of and preparation for the potential risks associated with climate change. The task force included members from DEP's multiple internal bureaus, the Mayor's Office, and other New York City agencies, Columbia University, and engineering consultants.

The task force conducted extensive internal interviews to identify potential impacts to DEP systems, frequently met with key science advisers, initiated a preliminary inventory of DEP's own greenhouse gas emissions, and participated in several major national and international conferences to share ideas and establish active partnerships with other municipalities and utilities around the world. DEP's *Climate Change Action Plan*, set forth by the task force in a May 2008 report, developed five tasks:

1. Work with climate scientists to improve regional climate change projections
2. Enhance DEP's understanding of the potential impacts of climate change on the department's operations
3. Determine and implement appropriate adaptations to DEP's water systems
4. Inventory and manage greenhouse gas emissions
5. Improve communications and tracking mechanisms

The action plan included two primary priorities for further study. The first priority was to conduct a phased, integrated modeling project to quantify and provide a more comprehensive understanding of the potential impacts of climate change on drinking water quality, supply, and demand. The second priority was to conduct an integrated modeling, inundation, flood mapping, and cost-benefit analysis project to quantify and provide a more comprehensive understanding of the potential impacts of climate change on drainage, wastewater treatment processes and infrastructure, and harbor water quality. These projects were completed in October 2013 and presented in two separate reports: the *Climate Change Integrated Modeling Project: Phase I* report and the *NYC Wastewater Resiliency Plan*. DEP's overall adaptation strategy was also captured in the "Water and Wastewater" chapter of *PlaNYC: A Stronger, More Resilient New York*, released in June 2013.

The Climate Change Integrated Modeling Project (CCIMP) evaluates potential impacts on New York City's water supply system using downscaled climate change projections coupled with DEP's watershed and reservoir modeling tools. The study found that the timing of spring snowmelt is projected to shift, from a distinct peak in late March and April to a more even distribution throughout the winter and fall, due to increased temperatures causing less precipitation to fall as snow, decreased snow accumulation, and earlier snowmelt. The shifting seasonal pattern in streamflow could lead to increased turbidity (a measure of sediment) in the fall and winter (Figure 15.3). Turbidity has presented a challenge in recent years, particularly in 2011 when back-to-back Tropical Storms Irene and Lee produced elevated turbidity and high bacteria counts in several of the city's reservoirs. As a result, special treatment continued for almost 9 months, the longest such treatment period ever recorded. However, with treatment and operational measures, DEP ensured that the drinking water delivered to the public remained in compliance and safe for consumption. As climate shifts, DEP will continue to track observed and projected changes to reservoir water quality, modifying reservoir management accordingly.

Under future climate scenarios (SRES A1B, A2, B1), greater streamflow may lead to higher reservoir turbidity. Turbidity over 5 NTU requires operational adjustments and, if necessary, special treatment.

The *NYC Wastewater Resiliency Plan* was born out of a pilot study initiated in February 2011 to assess potential impacts of climate change on drainage and wastewater treatment. The original *Climate Risk Assessment and Adaptation Study* was nearing completion when Superstorm Sandy hit in October 2012. Sandy caused more than $95 million in damage at 10 wastewater treatment plants and 42 pumping stations, and caused millions of gallons of untreated and partially treated wastewater to spill into New York City waterways. DEP quickly reacted to repair damage, and expanded the pilot study to include all at-risk wastewater treatment plants and pumping stations. The asset-by-asset analysis of the risks from storm surge considered the existing 100-year floodplain with a projected 76 centimeters of sea level rise by the

2050s, representing the 90th percentile of model projections from the New York City Panel on Climate Change's report *Climate Risk Information 2013*. The Resiliency Plan found that more than $1 billion in wastewater infrastructure is at risk, including assets at all 14 treatment plants and 58 pumping stations. The plan recommends $315 million in cost-effective upgrades at these facilities to protect valuable equipment and minimize disruptions to critical services during future storms. It is estimated that this investment in protective measures will help protect this infrastructure from over $2 billion in repeated flooding losses over a period of 50 years (Figure 15.3). DEP is currently implementing these improvements in new capital projects and is seeking funding to retrofit additional facilities.

DEP is also implementing additional resiliency measures that include improving energy reliability for water and wastewater facilities, improving and expanding drainage infrastructure, and promoting additional redundancy and flexibility of the water supply system. The department is monitoring observed and projected climate change to ensure adjustments are made to the system as needed, with help from the New York City Panel on Climate Change that includes many of the climate scientists that began working with DEP in 2003. Building upon the work advanced by its Climate Change Program, DEP continues to work across multiple city agencies to implement the cross-cutting initiatives presented in *A Stronger, More Resilient New York*, and as part of the Water Utility Climate Alliance (WUCA) to understand and advance the state-of-the-art in climate resiliency. New York City's and WUCA's efforts have been recognized as part of a collective of exemplary U.S. organizations that are working to overcome the barriers to climate change adaption.

## List of Terms

**Incident sunlight:** Light from the sun that falls directly on a defined area for a set length of time. Often expressed in a standard measure of intensity such as candelas.

**Urban heat island (UHI):** A city or metropolitan area that is measurably warmer than nearby rural areas due to human activities. Modification of land surfaces is the leading cause of urban heat islands.

# 16

## Rio de Janeiro

With a population of just over 10 million, Rio de Janeiro is the second most populous city in Brazil after São Paulo. However, Rio's rate of urbanization (97%) is greater (Confalonieri et al., 2009). The challenges Rio faces are unregulated settlements in hazardous areas where the urban poor live, industrial waste, and sewage removal (De Sherbinin et al., 2007).

### Hazard Risks

Rio de Janeiro faces high environmental risk levels. Hazard risk represents a cumulative score based on risk of cyclones, flooding, landslides, and drought. Rio's coastal location and population density heighten its vulnerability, making its dramatic topography more prone to certain types of hazards (De Sherbinin et al., 2007).

### Urban Poverty and Vulnerability to Climate Change

Dense concentrations of Latin America's urban poor dwell in packed, unhealthy shantytowns, making the populations highly vulnerable to extreme weather events. In Rio, the most vulnerable low-income people live in slums (favelas) that are often located on hills within the city. Favelas are usually built with lower standards of security and located in riskier areas. Landslides and floods often take people's lives. Sea elevation in the next decades might affect people living in some of these areas, which will demand adaptation of infrastructure (Paes and Rosa, 2013).

Favelas are unregulated settlements with few, if any services. These slums are populated by more than 1.1 million people in the city of Rio and they sprawl over the hills and slopes of the Tijuca mountain range (De Sherbinin et al., 2007).

Although living conditions have improved through the years, much still needs to be done. Sidewalks deteriorate over time, which increases runoff in the rainy season. Water runs down from the mountains to flood the lowlands, creating a sanitary problem and spreading certain diseases related to water contamination. Wastes accumulate when they are not collected and constitute a special concern for environmental health in Rio. Furthermore, disposal and treatment of waste can produce emissions of several greenhouse gases, which contribute to climate change, creating a rebound effect (De Sherbinin et al., 2007).

## Epidemics after Natural Disasters

The incidence of disease is affected by climate. Nevertheless, migration, sociodemographic influences, nutrition, and drug resistance, as well as environmental influences, are other factors that contribute to health issues. Contaminated water and sewage as a result of climate change are causing illnesses. After a flood-related outbreak of leptospirosis in Brazil in 1996, spatial analysis indicated that incidence rates of leptospirosis, which is an epidemic-prone zoonotic bacterial disease, had drastically increased inside the flood-prone areas of Rio de Janeiro, which are mostly located in the slums. This disease is transmitted by contact with contaminated water (Watson et al., 2007).

## Reducing Rio's Population Vulnerabilities

Rio de Janeiro hosted both the World Cup in 2014 and the 2016 Summer Olympics, presenting an intrinsically new challenge for the city (Laffiteau, 2012). Mitigating the adverse impacts of air pollution, urban transportation, and GHG emissions that contributes to climate change will require the government of Rio de Janeiro to take the lead in deterring the detrimental effects of harmful pollutants, flooding, and some other drastic environmental risks that could jeopardize the sporting event.

Some extreme weather and climate events affect the built, natural, and social infrastructure of cities. It could also affect the city's slum dwellers due to the consequences of flooding, high temperatures, and contaminated air (Laffiteau, 2012).

## Climate-Proofing Rio de Janeiro

Many floods and landslides have demonstrated the vulnerabilities of Rio de Janeiro and its high exposure to risk from a changing climate. In April

2010, a catastrophic weather event left approximately 200 people dead due to floods and landslides. The event also made several thousands of people homeless, and in early 2011, 450 residents died in mudslides in the state of Rio de Janeiro (De Sherbinin et al., 2007).

The city of Rio de Janeiro holds 5 million people, and the metropolitan region has approximately 11 million. The topography of the city makes it susceptible to landslides and flooding in low-lying areas. The diminishing areas of Atlantic rainforest expose the vulnerabilities of the thin soil, which has been stripped away from many hillsides. This effect, along with granite and gneiss bedrock that have been left exposed to weathering, creates landslides, decomposition, and erosion. Flooding is frequent during the rainy season, between January and March, because of the topography, with low-lying areas and a lack of drainage that makes the situation more susceptible to those weather events (De Sherbinin et al., 2007).

## Population Density

More than one million people live in favelas (approximately 20% of the municipalities' population). The population density of Rio de Janeiro is 4,640 persons per sq km, but densities in the smaller administrative units of the metro area are between 8,000 and 12,000 persons per sq km (De Sherbinin et al., 2007).

As the large and poor migrant population coming from Brazil's arid northeast region settles in Rio, government efforts are in place to deter these new migrants trying to move up hillsides to search for new land. The government regulations to control the favelas and to restrict building in hazard-prone areas are initiatives that help to improve access to basic infrastructure and avoid risks (De Sherbinin et al., 2007).

## Extreme Precipitation Events

Climate change is expected to increase precipitation and exacerbate the population health risks (De Sherbinin et al., 2007). Extreme and more frequent precipitation events have serious repercussions for the rates of mortality and morbidity. Moreover, floods and storms also raise the rate of deaths and nonfatal injuries, and diarrheal and vector-borne diseases (Nerlander et al., 2009).

Climatic profile, the continental size of Brazil's territory, geographical features, and social and economic issues affecting large population groups are characteristics of the country (Confalonieri et al., 2009).

Rio de Janeiro confronts unique health risks. The most serious natural event is heavy rainfall that provokes landslides and flooding. These events occur more frequently in the summertime. The city is acting to adapt and better respond to that risk with a crisis management system. The Center of Operations of the City of Rio de Janeiro is a forecasting system that is based on a unified mathematical model of Rio that pulls data from the river basin, topographic surveys, the municipality's historical rainfall logs, and radar feeds. The system predicts rain and possible flash floods and has also begun to evaluate the effects of weather incidents on other city situations, such as city traffic or power outages. That allows city leaders to develop reaction protocols that can help to reduce damage and to protect lives. The Center of Operations also assists in organizing information to support decision-making (Paes and Rosa, 2013).

## Tropical Cyclones: A New Reality

Tropical cyclones have never occurred in the vicinity of Rio de Janeiro, and the city has not been affected by an extreme tropical storm in several decades. (However, the state of Santa Catarina in southern Brazil was hit by a hurricane in March 2004.) In addition, Rio receives a strong El Niño-Southern Oscillation (ENSO) signal as well as higher precipitations during El Niño years. In 1988, Rio de Janeiro experienced extreme floods as the result of two intense periods of rainfall that caused 480 millimeters of rain, which equals one-third of the total rainfall in a year (De Sherbinin et al., 2007).

## Vulnerabilities

Vulnerabilities can be highlighted in different categories emphasizing the factors that determine a poor response to climate change for the city of Rio de Janeiro and the country as well. These vulnerabilities are differentiated as follows:

- Health impacts such as climate-sensitive infectious diseases of public concern
- Social impacts, which are factors that determine an ineffective reaction to climate change risks

Figure 16.1 presents the vulnerabilities of climate change impacts on health. Policy-making processes could include this vulnerability description

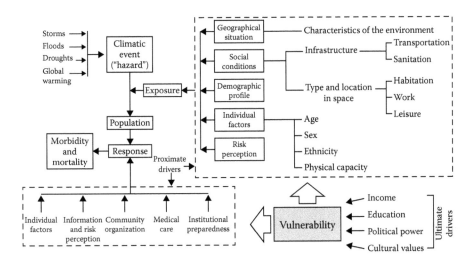

**FIGURE 16.1**
Conceptual model for vulnerability to health impacts of climate. (From Confalonieri, U. et al., 2009, Public health vulnerability to climate change in Brazil, *Climate Research*, 40(1), 175–186.)

aiming at the implementation of climate change adaptation. Proper policies for adaptation may reduce current vulnerabilities and improve the execution of climate change adaptation strategies (Confalonieri et al., 2009). The methodological–theoretical framework was based on a general "exposure–response" model. Identification of proximate drivers of vulnerabilities, ranging from individual traits such as age, sexual category, and physical aptitude to social–environmental characteristics such as geographical features, was included. Furthermore, the magnitude and timing of human exposures to climatic hazards, and the capacity and effectiveness of the responses to the impacts resulting from exposures, were modulated to incorporate all aspects of vulnerability to health impacts of climate. Structural characteristics such as education, income, governance, and political power, which were called "primary" or "ultimate" drivers of vulnerability, are shown on the bottom right of Figure 16.1, as well as the ultimate consequences of climate change impacts, which are morbidity and mortality as a result of social incapability and response to climate change impacts (Confalonieri et al., 2009).

## Urban Areas and Poverty: A Closer Look

Urban areas in Rio de Janeiro have been expanding for decades mainly in vacant lands that are not suited for occupation. These urban areas are composed of wetlands, steep hillsides, outcrops, rocky shores, estuarine channels, rivers, and forest remnants. The poor are settling in hazardous environments that affect their quality of life and, subsequently, their health. Considering that many of Rio de Janeiro's urban residents live under precarious conditions that are particularly concentrated in popular areas and slums, they share characteristics such as scarcity of water supply, lack of sewerage and street paving, illegal occupations, and unhealthy housing conditions, among others. These conditions frame the social, political, and economic inequality of the urban poor and highlight their vulnerable situation. Moreover, the equitable distribution of resources poses a significant challenge in addressing climate change effects on the urban poor (Pelling, 2011).

## Rise in Sea Level

The increase in the mean level of the ocean is a primary concern for coastal cities as climate change magnifies the risks associated with its effects and consequences. Average sea levels have been increasing in the last few decades with considerable regional variation. By the year 2080 sea-level rise will affect 5 times as many coastal residents as it did in 1990 (United Nations Habitat, 2011b).

The effective response to new meteorological and oceanographic conditions is crucial for coastal residents to counter the effects of climate change. The urban poor may not respond adequately to those conditions because of Rio de Janeiro's diversified and extremely modified coastal geomorphology. These coastal characteristics will adversely affect the response of Rio's population to sea-level rise (Pelling, 2011).

## Special Characteristics of Coastal Cities

The characteristics of coastal cities and their vulnerabilities to climate change impacts are highlighted:

- Low elevation—The majority of coastal cities are located at or near sea level. Coastal flooding represents a threat due to changes in sea level. Tidal-wave effects, cyclones, or frontal systems are also climate events exacerbated by climate change (Pelling, 2011).
- Topography—Some coastal cities are surrounded by mountains/topographic barriers. These geographic obstacles serve to increase precipitation (Pelling, 2011).

- Land use—Megacities serve as ports and points of trade, locally and internationally. Industrial activities are also part of their economic activity and result in GHG emissions with significant health consequences for local and regional populations, and deteriorating air quality (Pelling, 2011).
- Sea/land breezes—Water and land nearness creates daytime sea breezes and land breezes (onshore and offshore flows of air, respectively). This effect concentrates pollutants and recirculates them across the coast, with significant consequences for the health of coastal residents. The health implication is worse at night, as the urban boundary layer decreases and magnifies the harmful health effects (Pelling, 2011).
- Population density—The combination of scarcity of land availability and large populations creates an elevated division of impervious spaces/low fraction of greenspace, as seen in Rio's slums. The urban effects, particularly urban warming, are magnified due to energy flux partitioning-latent heat fluxes (which is the evapotranspiration rate; links between the properties of the urban environment and the exchange to the atmosphere). The evapotranspiration rates are suppressed, while sensible and storage heat fluxes are enhanced, creating greater heating of the air and substrate, intensifying urban effects (Pelling, 2011).

Rising sea levels will magnify the highlighted areas under risk, and the increase in frequency and intensity of storms associated with the elevation in the ocean temperature will exacerbate the problem in urban areas and particularly in large coastal cities such as Rio de Janeiro (Muehe, 2010).

## Management Plans to Address Climate Change

The new master plan for the city of Rio de Janeiro incorporates climate change mitigation and adaptation among the environmental guidelines already established. Furthermore, there are recent enactments such as the State Law on Climate Change (State Law No. 5690, of April 14, 2010) and the Municipal Law on Climate Change and Sustainable Development for reducing greenhouse gas emissions (Law No. 5248, enacted on January 28, 2011) (Barata, 2013).

The municipal strategic plan developed for the period 2009 to 2012 addresses the following initiatives considering climate change. The main objective is to implement a greenhouse gas emissions mitigation strategy (Barata, 2013):

- Implement new solutions for waste disposal in the city, closing the landfill of Gramacho. (Rio's Jardim Gramacho landfill, Latin America's largest, covers a landmass of 1.3 million square meters. This landfill is considered responsible for delivering tons of contaminants into Guanabara Bay.) It would enable an 8% reduction in emissions of greenhouse gases by the end of 2012, considering as reference the inventory of 2005 emissions.
- Expand the network of cycling in the city of Rio de Janeiro.
- Expand the public transportation system (BRT and subway) to reduce traffic congestion.

Furthermore, city leaders are working on a broad and deep adaptation plan for Rio de Janeiro. There is an established risk analysis strategy for the most vulnerable settlements (favelas) that has helped implement interventions when needed. Gradually, the city is improving that analysis with more information about the exposure to hazards of its population and their precarious housing infrastructure. The protocol of interventions and crisis management plans are all based on that information (Paes and Rosa, 2013). For the 2016 Olympic and Paralympic Games, Rio followed the model used for the London 2012 Olympic Games to understand and measure its carbon footprint. Initially, the city's plan is to anticipate any carbon impacts so they can be avoided, reduced, or compensated. Nevertheless, technical parameters have been adapted to the Brazilian reality and updated according to recent scientific developments on the subject. This methodology refers to various international standards, particularly the GHG protocol, and it is based on the best data, assumptions, and estimations available. Rio 2016 emission reduction efforts focus on

- Avoiding emissions through careful planning
- Reducing embodied carbon in materials through smart design and sustainable purchasing
- Substituting fossil fuels for renewable and alternative fuels (Rio 2016 Organising Committee for the Olympic and Paralympic Games, 2014)

## Municipal Services and Climate Change

Transportation is the main source of carbon emissions in Rio de Janeiro, according to the most recent greenhouse gas inventory of its 2005 emissions.

It accounts for 48% of emissions within the city. All policies and public concessions for transportation services such as bus, bus rapid transit (BRT) services, and urban bike paths are highly essential for the implementation of plans that promote a sustainable living and tackle climate change. The mayor of Rio de Janeiro, Eduardo Paes, is working on greener solutions for the transportation sector. The city is currently testing several technologies such as hybrid, electric, and biofuel to address its greenhouse gas emissions impact, emphasizing many innovative alternatives for mobility with a greener approach (Paes and Rosa, 2013).

# 17

## Beijing

China's remarkable economic growth over the last few decades, and the increases in production from many sectors of society, such as trade and manufacturing, is associated with rapid increases in the use of fossil fuels, the primary source of carbon dioxide. Economic activity and energy demand keep growing at an accelerated rate, which carries environmental consequences and poses serious challenges. For example, if China continues to consume energy from coal at its present pace, by the year 2020 China's average per capita energy consumption will match today's global average and will account for almost one-third of the world's total greenhouse gas (GHG) emissions (Hongyuan, 2008). Climate change combined with the current conditions presents a complicated situation for key cities in the country. Local governments need to address these issues because the environmental degradation caused by these factors will ultimately affect China's economic development.

## City of Ember

One of the 20 largest cities in the world, Beijing is also one of its most heavily polluted. Before the 2008 Olympic Games, Beijing was engaged in a series of actions to achieve a "Green Olympics," resulting in investments in alternative-fuel buses and taxis and improved auto emission standards. These accomplishments helped to shape adaptation and mitigation strategies in response to climate change. Because Beijing is an important political and economic center, it serves as a useful comparison city for case studies of other global cities that are creating their own adaptation and mitigation strategies to counter the effects of climate change (Zhao, 2011).

## Climate Change and Beijing's Population

Beijing is located near the Bohai Sea, and it covers an area of 16,807.8 km². The population of this megacity is 15.4 million, with 12.9 million living in urban areas and 2.5 million living in the hinterlands. The immigrant population is estimated at 3.57 million people. The impacts of climate change on the city's population concern scientists and health officials. The interactions of rapid urbanization and the effects of climate heighten these concerns. In fact, Beijing residents are already experiencing increased frequency of heat waves in the center of the city and with less cooling at night (Qi et al., 2007).

## Urban Heat Island Effect in Beijing

The accelerated urbanization of developing countries makes planning and implementing strategies to mitigate the effects of a changing climate an important topic for megacities. Considering the environmental and demographic issues in and around Beijing, adding to the knowledge of climate changes can help regulate urban ecosystems (Liu et al., 2009). Beijing's rapid development carries a series of consequences affecting its residents, and a series of studies has observed this phenomenon. The temperature in urban areas increased compared to rural regions from 1961 to 2008 (Biao, 2011).

The detrimental impact of the urban heat island in Beijing affects human health, adds to the number of heat-related illnesses, and deteriorates the air quality. Observations contained in an evaluation conducted by Chengshan Wang and Xiumian Hu from July to August (1993–2003) concluded that the urban heat island effect has increased in intensity every year. A heat wave affected the Beijing population from July 7 to July 15, 2003. The temperature during that period reached 35°C and emergency hospital admission visits increased 30% as a result of heat stroke and heat exhaustion. Additionally, 4,500 admissions to children's hospitals were reported as a consequence of fever and intestinal diseases (Qi et al., 2007).

## Air Quality in the Megacity

Approximately 70% of the air quality in China's urban areas does not meet national ambient air quality standards. Seventy-five percent of urban residents are exposed to air pollution because vehicle ownership has increased dramatically in Beijing from 0.5 million in 1990 to 2 million in 2002. Air pollution from coal smoke and automotive exhaust fumes, including 80% CO and 40% $NO_x$, are the main sources of industrial and domestic air pollution. In addition, pathogenic microbes, heavy metals, and toxic organics, which are a consequence of rapid urban expansion, are deteriorating the air quality (Qi et al., 2007). The effects of air pollution, especially aerosols (particulate matter, known as PM), have been indicated to be leading to global warming and consequently detrimentally impacting the health of Beijing's population (Raes and Seinfeld, 2009).

---

## Temperatures in Beijing

Based on the observed evidence of climate change, China's annual average air temperatures have increased by 0.5°C to 0.8°C during the past 100 years, and Beijing is now experiencing fluctuating increases in temperature that have risen by 1.7°C, from 11.6°C in 1978 to 13.3°C in 2009. The projections for 2020 are an increase of 1.3°C to 2.1°C and 2.3°C to 3.3°C in 2050 in comparison with 2000 temperatures (Biao, 2011).

---

## High Temperatures and Mortality in Vulnerable Populations

Research conducted in greater Beijing has established the exposure–response relationship between temperatures and residents' death rate. Assessing the risks of premature death related to higher temperatures contributes effectively to the development of new policies addressing the impacts of climate change on health. For this particular study, Beijing was divided into three regions: urban, inner suburban, and outer suburban areas, according to the classification of the Beijing Statistical Yearbook (National Bureau of Statistics of China, 2008). The main reason for this separation for the study is that health risks vary according to the areas chosen and many differences exist in the population of each region (Li et al., 2012).

The urban area in the study included the Dongcheng, Xicheng, Chongwen, and Xuanwu districts. The inner suburban area encompassed the Chaoyang, Fengtai, Shijingshan, and Haidian districts. Finally, the outer suburban area comprised the Fangshan, Tongzhou, Shunyi, Changping, Daxing, Mentougou, Pinggu, and Huairou districts, along with Miyun and Yanqing counties. The method used for the study was developed using a Poisson generalized linear regression model with Beijing mortality and temperature data from October 1, 2006, to September 30, 2008. This statistical model calculates the exposure–response relationship for temperature and mortality in the central city, and the inner suburban and outer suburban regions. A health risk model was also used to evaluate heat-related premature death risks in the summer of 2009 from June to August. The findings have shown that the highest mortality rate was recorded in the outer suburbs. Residents from the central city had a midrange risk, and the inner suburbs presented the lowest risk among its inhabitants (Li et al., 2012).

The number of deaths according to the risk assessment performed was 1,581. The highest number of premature deaths was recorded in the Chaoyang and Haidian districts. The Fangshan, Fengtai, Daxing, and Tongzhou districts showed a midrange rate in mortality. Considering the preceding facts from the statistical model used, premature death and higher temperatures are closely related in the city of Beijing. The vulnerability of the population to temperatures was highlighted in the city and outer suburban area. Table 17.1 shows the daily average mortality and temperature in Beijing from October 1, 2006, to September 30, 2008 (Li et al., 2012).

The numbers of deaths in the summer of 2009 in Beijing are shown in Figure 17.1 according to region. The central urban and outer suburban areas have experienced a greater effect of temperature on mortality. The different

**TABLE 17.1**

Daily Average Mortality and Temperature

|  | N | $x \pm s$ | Min. | Max. |
|---|---|---|---|---|
| Daily average |  |  |  |  |
| Mortality (no./d) | 731 | 168.3 ± 34.5 | 53.0 | 273.0 |
| Temperature (°C) | 731 | 13.8 ± 10.8 | −6.8 | 30.7 |

*Source:* Li, T. T. et al., 2012, Assessing heat-related mortality risks in Beijing, China, *Biomedical and Environmental Sciences*, 25(4), 458–464.

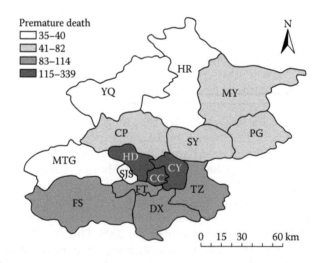

**FIGURE 17.1**
Temperature-related premature deaths. Central City (CC), Dongcheng (DC), Xicheng (XC), Chongwen (CW), and Xuanwu (XW). Chaoyang (CY), Fengtai (FT), Shijingshan (SJS), Haidian (HD), Fangshan (FS), Tongzhou (TZ), Shunyi (SY), Changping (CP), Daxing (DX), Mentougou (MTG), Pinggu (PG), Huairou (HR), Miyun (MY), and Yanqing (YQ). (From Li, T. T. et al., 2012, Assessing heat-related mortality risks in Beijing, China, *Biomedical and Environmental Sciences*, 25(4), 458–464.)

vulnerabilities of the population to the temperatures' effect in relation to the areas of study are highlighted in Figure 17.1 (Li et al., 2012).

Economic status also influences the vulnerability of the residents. Low-income populations are more susceptible to the impacts of climate change and drastic weather events such as heat waves. Socioeconomic levels dictate the degree of risk exposure. High population density is concentrated in central Beijing so it has more residents with greater vulnerabilities to heat than other parts of the city. In addition, Beijing residents from the outer suburbs are relatively low-income and are also vulnerable to climate-change impacts on health (Li et al., 2012).

## Water Scarcity and Climate Change in Beijing

Beijing has a warm temperature that is described as a continental monsoon climate. Observed changes in Beijing's climate include an increase in annual average temperatures, a probable decrease in rainfall and runoff average, and more heat waves. Therefore, climate change actions should not be taken in isolation but rather in conjunction with other pertinent issues that affect the city's growth and development (Biao, 2011).

Beijing is located in a semiarid area and, as the current available water resources keeps decreasing at alarming rates, water scarcity is of urgent concern for its residents and government officials (Cui et al., 2009).

## Transportation Sector in Beijing

The global average temperature may be much more sensitive to greenhouse gases in the atmosphere than is commonly believed. According to NASA scientist Dr. James E. Hansen, to avoid severe climate change effects we may need to reduce the concentration of $CO_2$ in the atmosphere from today's 385 ppm to 350 ppm $CO_2$ by 2100, if not sooner (Ackerman et al., 2009). Road vehicle usage accounts for 74% of GHG emissions, and transport is responsible for 23% of the world's energy-related GHG emissions. These emissions have been increasing more rapidly in the transportation sector than in any other sector over the past decade. Furthermore, an accelerated increase of 80% is projected by the year 2030, if cities keep operating as they do today.

The concern comes from rapid urbanization and the accompanying increase in vehicle ownership, which grew 570% between 1990 and 2006. In China, 1% of the population drives 10% of its vehicles (Creutzig and He, 2008).

As Beijing faces rapid motorization, the city is also experiencing environmental and economic issues that it must urgently confront. The drastic increase in vehicles in Beijing poses an environmental hazard, and leaders need to implement a mitigation plan to confront this problem that affects the health of millions of residents. Means of transportation increased from 1.2 million in 2000 to 2.5 million in 2005 and to 3.5 million in 2008 in the city. The health of Beijing's inhabitants has deteriorated as air pollution gravely affects them and has become one of the leading causes of illness and death. The municipalities surrounding the city of Beijing are also affected. For example, in Tianjin, a municipality neighboring Beijing, air pollution was associated with $1.1 billion, or 3.7%, of GDP, in terms of health costs in 2003 (Creutzig and He, 2008).

Climate change mitigation strategies must highlight the transportation sector and its repercussions on environmental degradation and GHG emissions. External costs associated with motor vehicle use in Beijing, with the sum of global climate change externalities, equal approximately 7.5% to 15% of Beijing's GDP. These GDP percentages are contingent to suppositions of social costs that are converted into economic costs.

Evaluating and measuring social costs of climate change is challenging and can lead to uncertainty. However, if long-term costs and risks are internalized, the impact and consequences of climate change externalities can be parallel to congestion costs and air pollution (Creutzig and He, 2008).

---

## Greenhouse Gas Mitigation Initiatives

Energy efficiency has become an important factor for the Beijing Municipal Government to address. The city must take essential initiatives to change its energy structure, promote renewable energy, and advance education and awareness of energy conservation and environmental protection. Energy shortages is an issue city leaders are trying to battle, in addition to worsening air quality that sickens thousands of residents and even causes death. Furthermore, some of the aforementioned initiatives have also developed GHG reduction measures that are benefiting the health of inhabitants and encouraging lower-carbon economic development in Beijing (Zhao, 2011).

---

## Some Progress along the Way

Although Beijing shows an increase in the total amount of energy consumption and $CO_2$ emissions, which is related to its rapid economic growth, these have decreased per unit of GDP. The city has made great strides to increase energy efficiency and reduce energy intensity. These vital efforts put the city of Beijing in the lead of mitigation strategies among other major Chinese cities. It also achieved $CO_2$ reduction through an increase in forest coverage and wetlands and, between 2001 and 2006, all these activities accounted for 80 million tons of $CO_2$ reduction (Zhao, 2011).

# 18

## Los Angeles

Greater Los Angeles has a population of nearly 18 million residents. This megacity generates nearly $750 billion in economic activity every year (Hall et al., 2012a). The city of Los Angeles has a population of approximately 4 million people in diverse communities within a 469 square-mile area. Part of the Southern California economy, and the 14th largest in the world, Los Angeles is also one of the chief centers of global trade and has a well-established entertainment industry (City of Los Angeles, 2007).

Los Angeles County is a good representation of how large municipalities work within the United States' structural government. These cities are not a formal political designation structure as some other major cities are around the world. Instead, Los Angeles is a reflection of urban sprawl or the "suburbanization" of America. The division or fragmentation of government is evident and presents a challenge in dealing with local and global issues affecting Angelinos (as city residents are known). The blurred lines between political structures and urban boundaries make dealing with traffic congestion, air pollution, and water contamination difficult problems to confront. Inner cities have financial challenges in addressing climate change impacts on the poor, as well as other urgent issues such as crime, population density, and health risks (Schroeder and Bulkeley, 2009).

## Effect of Climate Change on Air Quality

Air quality and severe weather events are interrelated and highly sensitive to climate change. Also, high temperatures and surface ozone in polluted regions cause harm to the environment and magnify the response to climate changes.

The detrimental effects of climate change will be felt with unrelenting severity in polluted areas by 1 to 10 ppb (parts per billion) over the coming decades. The research for this book shows that urban areas and populations will be severely affected, as will cities with pollution occurrences such as Los Angeles, according to the general circulation model and chemical transport model (GCM–CTM) studies of the harm that climate change exacts on air quality. The results will affect concentrations of particulate matter (PM), such as from soot released into the air, according to the GCM–CTM studies by +0.1 to 1 µg m–3 over the coming decades. Because the county of Los Angeles and the state of California are regularly affected by wildfires, these uncontrolled fires in combustible vegetation could become an important source of PM (Jacob and Winner, 2009).

The American Cancer Society (ACS) has studied the association of particulate air pollution with mortality, using a cohort to assess the connection between the two. Several studies confirmed the connection between air pollution and mortality, both all-cause and cause-specific mortality. The ACS studies and the Harvard Six Cities Study (one of the most influential, innovative, and longest-running experiments concerning the health effects of air pollution in America) have provided the basis for many estimates of the number of deaths caused by air pollution. The U.S. Environmental Protection Agency's National Air Quality Standard for Fine Particles, for example, has used these studies for its estimates (Jerrett et al., 2005).

Exposure to PM2.5 (tiny particles or droplets in the air that are 2.5 microns or less in width) may be even larger than earlier studies suggested within city gradients, as the relation between PM exposure and hazardous health effects have been established. The study also found that the effects of PM on health are nearly 3 times greater within city gradients in comparison to communities (Jerrett et al., 2005).

Using the dynamical downscaling method for obtaining high-resolution climate or climate change information from relatively coarse-resolution global climate models, a study examined the climate and air quality response to local $CO_2$ over Los Angeles for 6 months. The results presented an increase of ozone and PM as a result of $CO_2$ emissions. In general, global climate models have a resolution of 150 to 300 km by 150 to 300 km and the Los Angeles domain resolution is the highest resolution that has been detected in this particular region using the dynamical downscaling system (Z. Zhao et al., 2011).

---

## The Poor in Urban Communities

Heat waves can have profound consequences for poor residents of Los Angeles, because they do not have the necessary resources to pay for air-conditioning and eventually will leave hot spots and migrate from central city areas affected by these conditions (O'Brien and Leichenko, 2000). Migration will be one of the consequences of anthropogenic climate change. Many of the urban poor will not be able to migrate far and, consequently, Los Angeles will experience a migratory shift between neighboring counties (Friel et al., 2011).

## The Poor and Their Vulnerabilities

The wealth of Los Angeles and its diverse economy are well known, as Los Angeles is supported by leading entertainment, business and finance, health care, sports, transportation, and high-technology companies. However, 15% of the county population is living at or below the federal poverty level, including more than 1 in 5 children. Los Angeles also has one of the largest populations of homeless, which amounts to approximately 80,000 people (Cousineau, 2009).

Moreover, the urban poor and, more specifically, minority groups are also highly vulnerable to climate change, heat waves, and drastic weather events in Los Angeles. Low-income urban neighborhoods and communities of color are particularly vulnerable to heat waves because they are segregated in the inner city, which is more likely to experience the heat island effect. These subgroups lack access to health care centers and immediate aid to confront climate changes affecting their health. Recent research conducted in four California urban areas presents a positive relationship between the concrete infrastructures, heat-trapping surfaces, and community poverty, and a negative relationship between the quantity of tree cover and the poverty level of the community involved in the study (Morello-Frosch et al., 2012). This representation shows the inequality of heat island exposure to low-income populations compared with higher-income populations. Furthermore, African Americans show a higher heat wave mortality rate than other demographic groups in Los Angeles, which is nearly twice the average (Morello-Frosch et al., 2012).

## Water: A Vital and Valuable Natural Resource

Climate change puts at risk water management operations and the existing infrastructures that ensure the reliability of supply. As the threat of consequences increases, existing practices for designing water-related infrastructure must be adjusted to adapt to the current conditions. Also, drinkable water standards need to be met, and increased turbidity might considerably raise the costs and challenges of treating water. Structural measures are necessary for climate change adaptation strategies as well as for forecasting/warning systems, insurance instruments, innovative ways to improve efficiency of water use (e.g., via demand management) and economic and fiscal frameworks that attack the issue, such as the making of laws, institutional change, and new policies (Kundzewicz et al., 2008).

The majority of the 4 million Angelinos depends on Eastern Sierra Nevada water sources. The large natural reservoir of snowpack and melt-water is stored and, subsequently, delivered to the city by the Los Angeles Aqueduct (LAA), which is 547 kilometers long. The storage capacity of the seven reservoirs surrounding the aqueduct is approximately 37,000 hectare meters (300,000 acre feet), and the availability of this water supply source is crucial to the well-being of Angelinos and the future of the city's economy. Several problems with water supply and the quality of drinkable water have created new pressures for city leaders to address this issue. In 2007, the driest year on record, water-limited Los Angeles faced the lowest snowpack ever reported in the Eastern Sierra, where the city historically receives the greatest share of its water supply.

Over 70% of annual runoff originates from snowmelt in urban centers throughout the western United States. One-sixth of the world's population relies on these layers of snow, which accumulate in certain geographic regions and high altitudes where the climate includes cold weather for extended periods during the year.

All these constraints on water became a high priority for Mayor Antonio Villaraigosa (who left office in 2013) and illustrate the urgency of this problem (Mills et al., 2012). Public health is in danger as water quality deteriorates and quantity is reduced. Sanitation systems also will cause environmental contamination if a climate change strategy for water supply is not achieved (World Health Organization, Department for International Development, 2009).

## Los Angeles's Pioneering Initiatives

The city of Los Angeles leads the way in analyzing climate change and its impacts. In the first of a series of groundbreaking studies, University of California, Los Angeles (UCLA) researchers, with financial support from the city in collaboration with the Los Angeles Regional Collaborative for Climate Action and Sustainability (LARC), modeled temperature change at high resolution (Pascual, 2012). The researchers used a resolution of 2 km as opposed to 200 km and conducted the study within a period of time that should be relevant to policy makers as they develop and implement appropriate measures. A suitable time frame for implementation follows a 30- to 50-year span (midcentury; the years 2041 to 2060). The study considered two different greenhouse gas emissions situations: business-as-usual (RCP 8.5) and an aggressive mitigation strategy (RCP 2.6). The IPCC 5th Assessment Report has established these new terms, and they symbolize the range of policy options (high and low) that are generally presented in concessions in an international context. The concentrations of greenhouse gases are 1200 ppm or 460 ppm of $CO_2$ equivalent concentration by 2100, in that order (Hall et al., 2012b).

Los Angeles will experience a drastic increase in temperatures between the years 2041 and 2060. The results also show the benefit of future-climate modeling on a regional scale, due to the city's complex climatic environment given its proximity to the ocean, mountainous surroundings, and dissimilarity in seasonality (Hall et al., 2012b).

Figure 18.1 shows temperature extremes in Los Angeles. The UCLA study used 19 global climate models downscaled for Los Angeles and employed a state-of-the-art regional modeling technique. The technique is called dynamical downscaling, and it utilizes intense computation at a high resolution to model the climate's physical processes. Porter Ranch and Sylmar will experience the highest level of temperature. Hollywood will also be exposed to a higher temperature, which will triple. A highlighted note to the study is that coastal areas will likely experience a warmer climate but the temperatures will be lower than the inland areas, where some, as noted on the map, will reach extremely higher temperatures (Hall et al., 2012c).

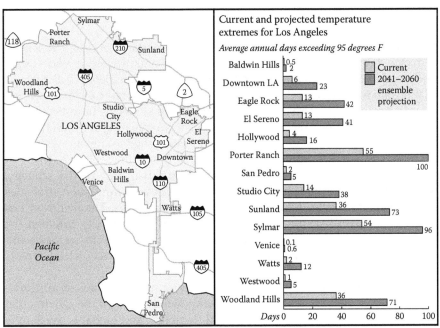

*Source: UCLA LARC study, 2012; chart based on the mean/average projected by the 18 climate models.*

**FIGURE 18.1**
Temperature extremes in Los Angeles. (Data from UCLA Professor Alex Hall's "Climate Change in the Los Angeles Region" project. For more information, see https://www.kcet.org /climate-change-la/temperature-study.)

## Mitigation Strategies

Los Angeles emits one-fifth of 1% of the world's carbon dioxide. Although it seems a minuscule number, it is equivalent to the carbon dioxide emissions of the entire country of Sweden. The plan, developed under Mayor Antonio Villaraigosa, combines different environmental initiatives with a goal of reducing greenhouse gas emissions 35% below 1990 levels by 2030 (Pascual, 2012). This ambitious plan includes urban planning strategies, which are crucial to achieve this goal and includes a series of initiatives to improve energy conservation, transition to renewable power sources, and change the ways residents commute to work and school. The plan encompasses actions to revamp the city to build more parks and open space, in addition to reducing municipal waste, improving water efficiency, and setting smart new standards for "green building" and land-use planning (Villaraigosa, 2007).

## Asking the Climate Question

The Center for Clean Air Policy (CCAP) has been working on issues associated with climate and air quality since 1985. A nonprofit organization, it works exclusively on these at local, national, and international levels. In 2006, CCAP started a relationship with Los Angeles county and city leaders to start the Urban Leaders Adaptation Initiative (Urban Leaders). Adaptation strategies have become an urgent agenda for Los Angeles and, as a result of participating in Urban Leaders, the city developed adaptation programs and established a Climate Adaptation Division within its Environmental Affairs Department (Lowe et al., 2009).

Microscale projects for adaptation plans are also part of the strategy, because a smaller scale approach does not conflict with mitigation strategies and creates efficiencies while implementing those plans. The health of Los Angeles residents became a priority for the mayor as the city initiated the Million Trees LA program (MTLA), a partnership between city agencies and not-for-profit community-based organizations to plant trees along streets and in parks (Lowe et al., 2009).

Furthermore, the Los Angeles Green Building program aims at minimizing the use of natural resources. The acknowledgment of the detrimental consequences of development on a local, regional, and global ecosystem is emphasized. The main areas the program addresses are

1. Site—Location, site planning, landscaping, storm water management, construction and demolition recycling
2. Water efficiency—Efficient fixtures, wastewater reuse, and efficient irrigation

3. Energy and atmosphere—Energy efficiency and clean/renewable energy
4. Materials and resources—Materials reuse, efficient building systems, and use of recycled and rapidly renewable materials
5. Indoor environmental quality—improved indoor air quality, increased natural lighting, and improved thermal comfort/control (Delgadillo, 2008)

## Moving Forward

The city of Los Angeles must make GHG mitigation strategies a top priority, considering the extraordinary volume of motor vehicle traffic within its territory. The Los Angeles County Metropolitan Transportation Authority (Metro) is developing sustainability strategies to emphasize transportation efficiency, access, and safety and operation excellence, while minimizing energy use, consumption, pollution, and waste. Because transportation is a focal issue for the city and its counties, the proposed plan includes those centered on Metro's vehicle fleet, buildings, and prospects to efficiently reduce vehicle-kilometers traveled (VKT) (Los Angeles County Metropolitan Transportation Authority, 2010).

Los Angeles is on track to implement a sustainability plan that will transform the city, because it is developing a strategy to add 60 new transit-oriented development areas alongside subway lines. These developments include dense housing, walkable streets, and mixed-used developments near transit options. The key goal is to help reduce GHG emissions by lessening the residents' dependence on motor vehicles. The city's dedication to sustainable operations is seen in the vehicles now on the road. More than half of the vehicles are hybrid or run on alternative fuels such as natural gas, which accounts for all the city's buses. What is more, air pollution from the Port of Los Angeles had been a problem for decades but now has been reduced by 45% (Los Angeles County Metropolitan Transportation Authority, 2010). The sustainability program under Metro follows the plan–do–check–act model that was established through Metro's Environmental Management System (EMS). EMS is an effective framework that can ensure compliance with environmental regulations and facilitate environmental stewardship. The Climate Action and Adaptation Plan conducted by the City of Los Angeles is a positive strategy to develop initiatives that will ensure a proactive approach toward adaptation and mitigation to lessen climate change impacts on the city and its population (Los Angeles County Metropolitan Transportation Authority, 2012).

## Metro's Greenhouse Gas Emissions Projections: 2020

Metro's internal emissions have increased substantially if compared with the 2010 report. By 2020, emissions will increase by 34,733 metric tons (MT), or 7%, to 511,220 MT. Most of the emissions increase will be from an expanded rail system, as rail emissions will account for the majority of total emissions in 2020 compared to 2010 emissions. Los Angeles's population is expected to keep increasing, and the Clean Burning Natural Gas (CNG) bus emissions are also expected to increase. Improvements in building energy efficiency and electricity supply that are less carbon intensive will provide a lesser impact on emissions emitted by the sector. Taking into consideration buses, rail, and vanpools, GHG will account for 95% of the Metro's transit service in 2020. On a positive note, GHG emissions per boarding will fall 4.4%, from 1.04 $MTCO_2e$ per thousand boardings to 0.99 $MTCO_2e$ per thousand boardings, even though annual boardings are projected to rise in 2020 to 516 million, up 12% from 2010 (Los Angeles County Metropolitan Transportation Authority, 2012).

## List of Terms

**Dynamical downscaling:** Downscaling is a term used in climate science to take information that is available at large scales to make predictions about events at local scales. Dynamical downscaling is one of two techniques used to make these projections. The other is statistical downscaling.

# 19

## London

According to the last census completed in March 2011, London has a population of 8,173,941 (Kaye, 2011). Within the boundaries of Greater London and counting the metropolitan area, the number of residents is between 12 million and 14 million. Greater London covers 609 square miles (1,579 square kilometers), and this fact makes the megacity the 37th largest urban area in the world. Furthermore, it is the most densely inhabited municipality in the European Union (Prasad et al., 2009).

## Coastal Flooding

London is in imminent danger of coastal flooding, and, as the climate changes and the population continues to grow, the threat will intensify. Londoners will feel the burden of the city's dense coastal population, poorly maintained housing, and other buildings that are not adapted to the threat. The intensity with which Londoners experience the consequences depends on what its leaders do now to minimize economic and environmental losses, in addition to saving lives. London is safeguarded by a system of fixed flood defenses, the mobile Thames Barrier, which protects its floodplain and provides a warning system that detects the high tides and storm surges moving up from the North Sea. This data is used to make decisions when to close the barrier. The vulnerability of residents to flooding is evident in the fact that 14% of Londoners are at risk of tidal and fluvial flooding. In addition, surface water flooding presents a threat to 3% of the population (Greater London Authority, 2014). Flooding has several causes, but the ones most applicable to London are as follows:

- River flooding that occurs when a watercourse cannot cope with the water draining into it from the surrounding land. This flooding can happen, for example, when heavy rain falls on an already water-logged catchment.

- Coastal flooding that results from a combination of high tides and stormy conditions. If low atmospheric pressure coincides with a high tide, a tidal surge may happen, which can cause serious flooding.
- Surface water flooding that occurs when heavy rainfall overwhelms the drainage capacity of the local area. It is difficult to predict and pinpoint much more so than river or coastal flooding.
- Sewer flooding that occurs when sewers are overwhelmed by heavy rainfall or when they become blocked. The likelihood of flooding depends on the capacity of the local sewerage system. Land and property can be flooded with water contaminated with raw sewage as a result. Rivers can also become polluted by sewer overflows (Environment Agency, 2009).

The megacity's older sewer system carries sewage and rainwater in the same pipe and consequently it causes spills and overflows. The result of the spills is a heightened risk of waterborne pathogens and a degradation of potable water; additionally inundations damage sewer and solid waste systems creating overflows and contaminating the area. As climate change impacts are increasingly more intense, London's plan for adaptation becomes a main priority for resilience.

## The Year 2100

Global sea-level rise due to climate change and local environmental changes will produce more frequent, longer-lasting, and more intense periods of flooding. The Thames Barrier will reach the end of its expected lifespan in 2030. Thus, in 2002, the Environment Agency established the Thames Estuary 2100 project to create a comprehensive risk management strategy to assess the vulnerabilities presented by tidal flooding, the impacts of heavy rain that leads to high river flows, and surface water flooding. For example, intense patterns of rainfall caused by climate change will lead to an increase of at least 20% in peak flows beyond 2050 and with a potential surge of 40% (Greater London Authority, 2014).

The Thames Estuary project considered the risks of tidal floods and the changes that might occur over time. Other considerations included the assessment of existing flood walls and barriers to determine which might become outdated, deteriorate, and unable to manage rising water levels. As a result, the need to prepare and protect the city and its residents forced leaders to plan for the future and recommend tangible actions to adapt to a changing estuary (Environment Agency, 2012).

## Preparing for Changing Climate

Adaptation is a dynamic process that requires continues improvement. How we measure risks today is unlikely to bring changes that will be entirely acceptable in the future because the level of protection needed will change as the climate does. Other factors that need consideration are the state of the city, its infrastructure, communications network, economic and social resilience, and transportation system. Drafting new policies, however, is insufficient to manage climate variability and change effects. Therefore, understanding and leading adaptation strategies will help to provide an acceptable level of protection in the future and integrate the strategy into the city's decision-making processes. These steps will result in better choices (Greater London Authority, 2011).

## London's Transport, Energy, and Waste Infrastructure and Initiatives

The transportation system is critical for London to function, and it is most likely that not all modes of transportation such as the underground network, train, buses, and so on, are going to be impacted by climatic impacts. Some types of transportation, however, can be affected more intensely than others. As Hurricane Sandy showed when it impacted New York City, an underground transportation network is the most vulnerable sector of the system. Flooding and overheating present a high risk to the London Underground lines since water will naturally flow to the lowest point and cooling the deep level can present difficulties for operating the system (Greater London Authority, 2011).

## Urban Accessibilities and Climate Change

The codependency of the urban form (the physical design of cities) and transport systems creates an environment with accessibility for all residents and brings economic and social benefits, as well as affects carbon emissions (Rode et al., 2014). This codependency is critical in order to evolve to a low-carbon economy (Hickman and Banister, 2014).

In London, the physical concentration of the city's residents plus its services, economic activity, and the state of the system network are critical to attaining accessibility. The result is the emergence of alternative urban development and accessibility pathways. A trend back toward more concentrated urban forms is also rising in many other European, as well as North American, cities (Rode et al., 2014). Susceptibility to environmental impacts is less when a city is redensified. As global population increases and migration to cities grows, city planners need to rethink how to design a city, restructure or retrofit the existing plans, and add new technology or features to older systems.

## Leading the Way to Carbon Dioxide Reductions

New technologies and initiatives to reduce greenhouse gases present cost-benefit opportunities and, therefore, preclude millions of premature deaths (Wilkinson et al., 2009). Minimizing $CO_2$ emission became a strategic plan for many cities around the world, and London has engaged in an ambitious plan to reduce emissions from ground-based transport.

The target is to reduce emissions 60% by 2025 from a 1990 base. The three main points of the plan are as follows:

- Improved operational efficiency as a priority to reduce $CO_2$ emissions.
- Low-carbon vehicles, technology, and energy as a driver of $CO_2$ reductions. Incentives and the role of stakeholders are required to efficiently develop and implement the plan.
- London has made a massive investment in low carbon models such as walking, cycling, and public transport. Because London sits on a river, water traffic and rail freight give the transportation sector an opportunity to lessen emissions and ease road and highway traffic congestion (Greater London Authority, 2011).

## London's Urban Heat Island

Cities such as London demonstrate a correlation between high-population density areas and heat-related mortality and morbidity. The heat wave that hit London in 2003 was the greatest in terms of deaths per capita, especially among the elderly. Several studies of heat wave causes and related deaths, including an analysis of the 1995 Chicago heat wave, revealed the risk for heat-related hospital admission was much higher in cities than in rural areas. Moreover, the same results were compared to suburban areas in Switzerland

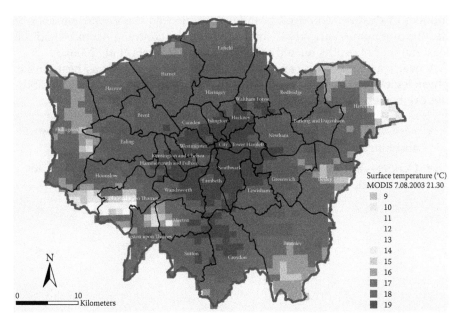

**FIGURE 19.1**
Surface heat island London, August 7, 2003, at 2100 h GMT (MODIS Image). Legend shows surface temperature in degrees Celsius. (From Wolf, T., and McGregor, G., 2013, The development of a heat wave vulnerability index for London, United Kingdom, *Weather and Climate Extremes*, 1, 59–68.)

in 2003. Therefore, numerous cases of London's heat wave during August 2003 could be the result of the urban heat island effect (Greater London Authority, 2006). Figure 19.1 illustrates the effect of the 2003 heat wave.

According to Wolf and McGregor (2013), London experienced severe heat wave events with harmful health repercussions in 1976 (MacFarlane, 1977), 1995 (Rooney et al., 1998), 2003 (Johnson et al., 2004), and 2006 (Health Protection Agency, 2006) with evidence emerging for distinct health effects of heat waves in July 2009 (Health Protection Agency, 2010) and June 2011 (Green et al., 2012). The evidence coming from these records clearly suggests a relationship between heat and mortality. The increase in mortality is most evident when London's average temperatures are above 19°C (66°F) (Hajat et al., 2002).

Short-term increases in mortality were reflected in the elderly during the first two weeks of August 2013. Deaths were evenly distributed at home, in hospitals, and in nursing or residential homes. It was also reported that with an increase by 42% during the heat wave, nursing homes experienced the greatest numbers of deaths. The most prominent disproportion was among the under-75 age group. Residential homes had an excess in mortality among this age group even though the confidence interval was wide due to the small number of deaths. Moreover, hospitals experienced the greater number of deaths; care homes carried 24% of deaths. On the contrary, hospices had the least number of deaths related to the heat wave (Kovats et al., 2006).

A heat wave's severity can lead to several illnesses such as heat stroke, physiological disruption, organ damage, and even death. Heat vulnerabilities can be described as follows:

- Age—The elderly suffer the greatest effects of heat waves in both absolute and relative terms.
- Gender—Socioeconomic factors make women more vulnerable to heat waves.

# 20

## Mumbai

Mumbai (formerly Bombay) is a cluster of islands, of which seven comprise the city proper and four its suburbs. It is considered the financial capital of India, and it is also home to two of India's major ports. This coastal megacity occupies an area of 468 sq km, and its width is 17 km east to west and 42 km north to south. It has a rich natural heritage, such as hills, lakes, coastal water, forests, and mangroves, alongside built areas. Because of its geographical location and its physical, economic, and social characteristics, Mumbai is vulnerable to the effect of climate variability and change, including storms, floods, and sea-level rises. Its growing population is the most important factor making it susceptible to climatic events. As a financial center, Mumbai attracts people searching for economic opportunities. This activity puts pressure on the operational efficiency of the infrastructure, civic amenities, and housing. Furthermore, this megacity is coping with traffic congestion, carbon emission from heavy vehicular circulation, solid waste issues, and the growth of illegal slum dwellings that are urgent problems exacerbating climate risks (Patankar et al., 2010).

For megacities, understanding the nature of their vulnerabilities is a challenge. It helps them to identify the risks and is crucial for an effective adaptation and mitigation plan. Mumbai's current population is about 21 million and its per capita income is a little more than 3 times that of India. The United Nations estimates that Mumbai's population will grow to about 27.8 million by 2030, with a 2015–2020 average annual population growth rate of 1.64% (United Nations, Department of Economic and Social Affairs, 2014).

## Vulnerabilities to Climate Risks

In July 2005, Mumbai was beset with India's heaviest rainstorm in a century. This catastrophic event killed nearly 600 people and paralyzed the city's operations from infrastructure, public health facilities, and transportation. It was one of the most devastating storms ever recorded in the country (Patil and Deepa, 2007).

Mumbai's built environment has problems directly related to its drainage system and its squatter communities. Water shortages, building collapses, and sewage disposal issues leave the city perilously exposed to severe weather events. Access to drinking water is limited, and sanitation is poor (De Sherbinin et al., 2007).

The old section of the city has a propensity to flooding, which leads to shutting down commuter rail lines. Subsequently, the flooding occurrences contribute to sea-level rise. In a national study of India's coastal zones according to their vulnerability, Mumbai was found very susceptible to harm.

Approximately 55% of Mumbai's population lives in slums, which are mostly located in low-lying coastal areas and along stream banks. Many of these precarious habitats are constructed with salvaged materials and are so deteriorated that they pose physical and health risks to the inhabitants. Mumbai's squatter communities are among the most densely populated districts in the world with a population as high as 94,000 people per square kilometer (De Sherbinin et al., 2007).

## Elements of Risk in India's Cities

Several elements contribute to Indian cities' vulnerabilities to severe weather events:

- Slum dwellers, squatters, and migrants live in traditional and informal settlements, which are often located in the most vulnerable locations
- Industrial and informal service sector workers, whose occupations place them at significant risk from natural hazards
- Buildings, especially traditional and informal housing that is especially vulnerable to wind, water, and geological hazards
- Industrial units, their in-house infrastructure, plant, machinery, and raw materials
- Lifeline public and private infrastructure, which includes roads, bridges, railways, ports, airports, and other transportation systems; water, sewage, and gas pipelines; drainage, flood, and coastal defense systems; power and telecommunication infrastructure; and critical social infrastructure such as hospitals, schools, fire and police stations, and first responder's infrastructure
- Ecosystems and the natural environment, especially wetlands, estuarine and coastal ecosystems, and surface and groundwater systems

Analyzing these elements that make a city susceptible to disaster leads to the conclusion that risk adaptation and mitigation measures must focus on particular urban populations and infrastructure. In the case of Mumbai's flooding in 2005, the city confronted a diverse array of constraints and challenges to efficiently respond to the disaster. Based on this experience, decentralized adaptive management strategies that engage political leaders and policy makers at all levels—from neighborhood to national—provide evidence of effectiveness to countervail the limitations and inefficacy of centralized top-down interventions (Revi, 2008).

The analysis of interventions shows that combining short-run priorities and long-run strategic actions under a universal framework can lead to a paradigm shift in current policies and management of the city. Also, private and public sector actions, urban development, and planning can support the strategy and minimize climate threats (Revi, 2008), because the process of risk management is ongoing, continual improvement. As new hazards arise and the city faces new challenges and changes to the environment, these must be discussed and dealt with.

## Urban Disasters: Flooding

In India, as in many developing countries, rainfall variability is central to issues of agricultural production, and, in some cases, it is at the core of the national economy. Consequently, extreme climate events impact the general economy and often affect the population in urban areas. The trends in rainfall in India have been studied since the last century, such as the long-term southwest monsoon/annual rainfall trends over India as a whole by Parthasarathy et al. (1994) and trend analysis for rainfall in Delhi and Mumbai by Rana et al. (2011), among others. Over the last 50 years, studies have shown a significant decrease in the frequency of moderate-to-heavy rainfall events over most parts of India. In some regions, however, extreme rainfall events are increasing at a rate of 100 millimeters per day (mm/day) as recorded by Goswami et al. (2006).

After the 2005 flood, economic losses and future scenarios were studied and presented by Ranger et al. (2010). The analysis presented a harsh picture for the future of Mumbai; the authors have stressed the need to evaluate and consider uncertainties in climate projections for adaptation planning in the city. Multiple available projections need to be considered like the global climate models and regional climate models, because a single scenario might not be a proper tool to obtain adequate information to apply to decision-making in adaptation planning (Rana et al., 2014).

As previously stated, floods in regions of India have grown in intensity and frequency. Rainfall has become more extreme, bringing densely populated cities like Mumbai to a standstill. Climate change has altered the frequency and intensity of these weather events in Mumbai, and 100-year floods might now happen within a decade or two and last longer. Also, during flood season, concentrated heavy showers affect the drainage system and paralyze the city's road and rail network, communications, and transportation systems (Alankar, 2015).

Research on climate change, funded by the Organisation for Economic Co-operation and Development (OECD) in the wake of the 2005 floods, concluded that the direct economic losses totaled US$2 billion and caused 500 fatalities (Alankar, 2015). At the same time, Mumbai's rapid urbanization has intensified threats to the urban population risks and dissuaded any adaptation effort.

## Response to Climatic Impacts

The Mumbai flooding led to a focus on disaster reduction and risk mitigation as an important factor to the adaptation agenda. In addition, an emphasis was placed on transferring key financial infrastructure and information technology (IT)-enabled services to other cities with lower risks of impact. The OECD research highlighted the weak response to climatic events by the administrators and proposed a series of activities to implement a strategy to cope efficiently with such impacts:

- Quantify immediate- to medium-term physical, economic, environmental and social outcomes resulting from selected weather events
- Characterize vulnerability by examining the trends in impact indicators
- Characterize responses in terms of costs, distributional effects, and efficiency
- Identify opportunities and means for incorporating climate risk into local and regional decision-making (Alankar, 2015)

## Delayed Response and Consequences of Strategic Unpreparedness

Mumbai's official reaction to the flooding is a reflection of how unprepared its response system was, the absence of a business continuity strategy, and consequently its increased susceptibility to future climatic events. Volunteer organizations took the lead in response to initiatives while the Brihan Mumbai Metropolitan Corporation (BMC) did respond much later with a disaster management plan, which uncovered the unpreparedness of the megacity's capabilities to cope with the magnitude of this event (Alankar, 2015).

As the floodwater rose in 2005, the heavy rains disrupted many sectors of the city:

- Transportation system—Train movement stopped, the roads were full of water, certain roads submerged, and the airport was unable to function for 2 days.
- Communication system—The communication network was disrupted as telephones and cell phones went down one by one. Had the network been operational and the public address system effective, many people would have received guidance and not put their lives in danger.
- Power supply—Due to submerged power stations and substations, the power supply in suburban areas got suspended.
- Water logging—About 22% of Mumbai's land was submerged in rainwater on July 26 and 27.
- Food and civil supplies—Commencing on the evening of July 26, the daily consumables could not reach the people.
- Rescue and relief measures—All efforts were made by rescue workers, but the rain totals made matters very difficult to handle. Measures were taken in terms of mosquito control and water purification. Additional physicians were employed to manage the increased patient load. Water was supplied through tankers. Many nongovernmental organizations were also active in rescue work (Envis Centre on Human Settlements, 2006).

## Leading Actions after the Event

The urban poor living in low-lying areas are exposed to flooding and other hazards, including inadequate storm water drainage. The city's storm water drainage (SWD) system consists of a hierarchical network, located mainly in the suburbs, which is a roadside surface drain 2,000 kilometers long and allows rainwater to run away freely from the road and highway. The storm water drainage system in the city is about 70 years old and able to handle rain of 25 mm per hour at low tide. If the intensity of rainfall is more than 25 mm and the city experiences hide tides, there is a possibility of water logging ("Storm Water Drainage," 2013).

Some city leaders recognized the urgency of the problem and took action. The Brihanmumbai Municipal Corporation (BMC) took the initiative and proposed the redevelopment of slums and villages on the coast under the Coastal Regulation Zones (CRZ) rules. However, there is no consideration of sea-level rise, which will affect these zones and their population. The actions of local people were the main reason the city returned to normal when the disaster management team and the corresponding authorities failed to act according to the urgency of the circumstances. Therefore, CRZ needs to recognize the significant role of local knowledge and strategies as past experiences abet in mapping the response to climate variability and change (Alankar, 2015).

## Desertification and Climate Change

Desertification, the process of land degradation in dry areas (Grainger et al., 2000), can be a result of climate change impacts and human activity. The understanding of such events is central to preventing desertification expansion. Desertification has detrimental economic, social, and environmental consequences, increasing human survival and developments risks (Xu et al., 2011).

The dependency of populations on natural resources, particularly in the developing world, can be severely affected by climate change, drought, and desertification. Hence, given the intensification of extreme weather events, it becomes crucial to examine local-level adaptation plans to cope resourcefully with those effects.

## Desertification and Migration

The link between climate change and international migration is problematic for many developing countries. However, research on the subject of internal population shifts suggests that is a significant factor in climate-induced migration. Climate-related desertification is causing migration in Mumbai as well as other cities like Delhi and Kolkata (DePaul, 2012). Research that investigated desertification and land degradation in India concluded that a 105.5 mega hectare area of the country, or 37% of the country's total geographic area, is undergoing these processes, and of this, the area undergoing desertification is 81.4 mega hectares (Ajai et al., 2009). These conclusions affirmed that desertification reduces job opportunities, causes mass migration from rural to urban centers, and presents global food security risks. Mumbai clearly presents a good example of climate-induced mass urbanization and resulting poverty. The movement of people to Mumbai is a result of internal migration coming from the south in addition to international migration from

**TABLE 20.1**

Climate-Induced Mass Urbanization and Urbanization of Poverty

| Condition | Result |
|---|---|
| Poorly designed and maintained buildings | Damage by extreme weather events |
| Transportation priorities | Limits disaster preparedness |
| Under severe climate events | Increases *pressure* on resources |
| Impoverishment (living in slums) | Slowing of local economy |
| Desertification | Migration, mainly from Bangladesh |
| Water scarcity | Worsening living conditions |

*Source:* DePaul, M., 2012, Climate change, migration, and megacities: Addressing the dual stresses of mass urbanization and climate vulnerability, *Paterson Review of International Affairs, 12,* 145–162.

surrounding countries. As climate change magnifies these issues, the city of Mumbai's vulnerabilities to such impacts will be exacerbated.

Mumbai was ill-equipped for the flooding of 2005. Contributing to its susceptibility were its topography, drainage system, and low river deltas. In addition, lax building code enforcement exposed the city's population to drowning, wall collapse, and various diseases. Those deaths were mostly among inhabitants living in slums (DePaul, 2012). These squatter communities comprise about half of Mumbai's population, and they are characterized by poor sanitation, precarious housing infrastructure, and the lack of basic services such as potable water (De Sherbinin et al., 2007). How mass urbanization under certain conditions lead to stress bundles is notorious in Mumbai. Table 20.1 lists conditions and their results.

As stated, desertification forces many migrants, particularly from Bangladesh but also locals from rural areas, to leave their homeland and seek refuge in the city. When the effects of desertification and water scarcity affect the population, their precarious socioeconomic conditions make them highly vulnerable to severe weather events because it is difficult for this segment of Indian society to mobilize and change their conditions.

## Built Environment

Poor living conditions and bad governance occur due to political and economic ineptitude, increasing the challenges of urban poverty. In this case, the foundation of the city's priorities is not directed toward human needs in an equitable and sustainable manner (United Nations Human Settlements Programme [UN-Habitat], 2013).

The city built environment has problems directly related to its drainage system and its squatter communities. Water shortages, building collapses, and sewage disposal issues leave Mumbai badly exposed to severe weather. Most of the old part of the city has a propensity to flooding, which leads to shutting down its commuter rail system. Subsequently, this flooding contributes to sea-level rise, and in a national study about India's coastal zones, according to their vulnerability, Mumbai was found highly susceptible to harm (De Sherbinin et al., 2007).

Approximately 55% of Mumbai's population lives in slums and mostly in low-lying coastal areas and along stream banks. Many such precarious habitats are in severe distress, and pose physical and health risks because they are constructed with salvaged materials. Mumbai squatter communities are among the most densely populated in the world with densities as high as 94,000 people per square kilometer. Nevertheless, accessibility to drinking water and sanitation is poor (De Sherbinin et al., 2007).

## Health and the Environment

Outbreaks of leptospirosis were reported in children living in precarious housing infrastructures after the flooding of 2000, 2001, and 2005 with an increased prevalence of eight-fold following the major flood event in July 2005 (Kovats and Akhtar, 2008). Leptospirosis is a bacterial disease caused by bacteria of the genus *Leptospira*. According to the Centers for Disease Control and Prevention, this disease can cause a broad range of symptoms. Without treatment, it can lead to kidney damage, meningitis (inflammation of the membrane around the brain and spinal cord), liver failure, respiratory distress, and even death (Centers for Disease Control and Prevention, 2014).

## Health Impacts of Air and Water Pollution

Mumbai has severe air and water pollution, according to the World Bank. About 75% of all sewage is unprocessed and discharged into local waterways and coastal waters. These actions cause environmental degradation and health risks to the population. Urban air pollution also affects the health of Mumbai's inhabitants as industrialization, increased vehicle use, and population growth has upped the health risks for greater numbers of residents. The World Health Organization (WHO) named Delhi as one of the top ten most polluted cities and Mumbai is next to it in air pollution levels (Shankar and Rao, 2002). An increased incidence of tuberculosis is an indication of air pollution health effects, as in the case of Mumbai. Moreover, ambient air concentrations of suspended particulate matter (SPM), nitrogen oxides

($NO_x$), and hydrocarbons have crossed the allowable limits of air pollution tolerance index, exacerbating health issues to Mumbai's urban residents (Yedla, 2003).

## Analysis of Mumbai's Climate-Disaster Resilience

Mumbai city is known for its financial and entertainment activity, but the megacity lacks the financial services, budget and subsidies, and savings and insurance to address the risks caused by severe weather impacts, climate variability, and change. Moreover, Mumbai institutions are fragile in terms of resilience. If disaster led to the city's collapse, the results would be violent disorder and interruption of basic services, causing direct and indirect harm to livelihoods. As the megacity's overall climate-disaster resilience is relatively low, stakeholders need to take a step forward to sustain efforts to strengthen physical, social, and institutional dimensions of resilience (Razafindrabe et al., 2009). The capacity of the megacity's infrastructure to maintain and provide services after a disaster is crucial for the health and survival of its population. Mumbai's experience with flooding makes its population aware of the consequences of such an event. Capitalizing on this awareness should help local government and civic societies to provide problem-solving strategies that build social resilience. Broader cooperation with other institutions and mainstreaming disaster risk reduction needs to be included in the megacity's development agendas as a primary objective for resilience (Razafindrabe et al., 2009).

# 21

## Lagos

Situated on the West African coast, Lagos is Africa's biggest city and also the fastest growing metropolis in the world. In 2011, the United Nations estimated its population at approximately 11.2 million. However, in its metropolitan area, this city is believed to have more than 20 million inhabitants (Shelley, 2015).

As the most urbanized city in Nigeria, Lagos holds most of the nation's wealth and economic activity. Floodplains define the urban area as a network of marshes, swamps, rivers, streams, and estuaries. The city receives large quantities of rainwater overflow and waste discharges from municipal and industrial sectors. The waste empties into the Lagos Lagoon (Diop, 2014).

A report by the Ministry of the Environments of Lagos State in 2012 shows a temperature increase of 0.04°C per year from now until the 2046–2065 time period. Moreover, annual rainfall, longer rainy seasons, heat waves, and severe weather events will be more frequent (Komolafe et al., 2014).

### Sea-Level Rise and Flooding

An assessment of 136 port-cities about flooding risks to exposed populations, Lagos ranked 30th in a 2005 scenario. Nevertheless, in a future climate scenario (2070s) it will rank 15th (Adelekan, 2010).

Several important factors contribute to this situation:

- African cities lack the infrastructure to withstand extreme weather conditions.
- Poor urban planning, together with other urban governance challenges, contributes toward placing African urban slum dwellers at highest risk.
- Poor urban planning, or the lack of planning for urban expansion, leads to new development in areas susceptible to flooding or in areas that should be left undeveloped (for instance, wetlands) because they are buffers against flooding risks.

The fact that low-income groups cannot find safer sites contributes to these increased risks (Adelekan, 2010).

## Slum Population Vulnerabilities

The slums in this megacity are densely populated with many people living in floating slums. Sea-level rise will displace this underserved population. The slums' flat landscape and low elevation, densely populated areas, obsolete infrastructure, and widespread poverty play a critical role in increasing the threats of natural disasters and climate change (Fifth Urban Research Symposium, 2009). The health implications of flooding include direct waterborne diseases like typhoid, cholera, pneumonia, diarrhea, and malaria. The increase of temperatures is known to intensify illnesses like meningitis, measles, and chicken pox, and it is suspected that new health risks like high blood pressure and dehydration in pregnancy can cause a higher morbidity rate (Omoruyi and Kunle, 2011).

## Air Pollution and Climate Change

The failure of local and global environmental policies aggravates air pollution problems in cities, which is the leading cause of premature death. It is also considered an accelerating factor that drives climate change and its effects in developed and developing countries. The detrimental effects are more pronounced in the cities of developing countries. Lagos, for example, has received insufficient funding for this type of issue. Lagos's leaders must face the effects of growing air pollution and its contributions to climate change. They must also understand the gravity of the public health risks and take action on a comprehensive health care strategy. Thus, they must implement corrective and adaptive measures to minimize the loss of lives and prevent the collapse of an overwhelmed health system (Komolafe et al., 2014).

## Risk Reduction Measures

Reducing risk involves a series of steps designed either to prevent hazards from occurring or to reduce the intensity and harshness of them. Involving different sectors of society in moderating vulnerabilities is an important step toward resilience. Creating an emergency plan to address safety and security, ensure an adequate health care response, and keep transportation and communication systems operating in stressful circumstances will raise awareness and help to prevent, mitigate, and prepare for emergencies and disasters. Mapping the flood prone areas is another major step to decreasing risk. It is crucial for megacities, like densely populated Lagos, to compare historical records of natural disasters to obtain indications of flood inundation areas, the physical coverage of the event, and the period of occurrence (Ladipo et al., 2011).

The high-water marks of earlier floods offer clear warnings of dangers associated with flooding. The land characteristics and tide levels of Lagos as a coastal megacity determine the submergence areas. Flood hazard mapping provides a clear picture of water flows. As in the case of Rio de Janeiro previously presented, land use control in Lagos has the potential to lessen the potential for loss of life and property damage during intense flooding. Land use is a particular determinant of flood-related deaths in coastal megacities.

Allocation of resources in areas where there is a settled urban population will reduce vulnerability. Areas predisposed to high floods must be restricted from major developments, and hospitals, schools, and other vital infrastructures must be built in safe areas.

The following are some suggestions for risk reduction measures:

- Create water holding areas such as ponds, lakes, or low-lying areas.
- Build structures engineered to withstand flood forces and seepage.
- Construct buildings in an elevated area. If necessary, build on stilts or platform.
- Reduce damage through flood control aims. This can be done by decreasing the amount of runoff with the help of reforestation (to increase absorption could be a mitigation strategy in certain areas), protection of vegetation, clearing of debris from streams and other water holding areas, conservation of ponds and lakes, and so on.
- Include levees, embankments, dams, and channel improvement in flood control plans. Dams can store water and can release water at a manageable rate. Flood proofing reduces the risk of damage; measures include use of sand bags to keep flood waters away and buildings should be constructed away from water bodies.
- Heed flood warnings. With the exception of flash floods, there is usually ample warning. Heavy precipitation warns of a coming river flood. High tides with high winds may indicate coastal flooding is imminent. Evacuation is possible with suitable monitoring and warning. Warning is important (Ladipo et al., 2011).

## Climatic Impacts and Transport System

Transportation networks are the hallmark of the growth and progress of civilization. Every nation's economy is measured by the efficacy and development of the transport infrastructure. When that infrastructure is damaged or outdated, the risk for the population is increased because response to emergencies may be delayed. Lack of mobility limits the capacity to manage climate vulnerability and change impacts.

Therefore, the city's transportation network is susceptible to climatic events. In fact, Lagos has an impaired ability to operate under normal conditions and puts millions at risk. Rising sea levels, an increasing frequency of hurricanes, floods impacting the city's facilities, and higher storm surges compound the problems. A disaster of any type keeps the transport system from working. Business continuity plans become crucial to resume critical operations after a disruption. Therefore, adequate attention must be given to adaptive resilience to ensure the economic stability and long-term development of Lagos, the economic pivot of Nigeria (Ede and Oshiga, 2014).

# Section V

# Prognosis for Change

# 22

## Risk Management, Strategy, and Leadership

### Leadership as the Starting Point of Strategy: Strategic Adaptive Cognition

The renowned writer, thinker, and lecturer Peter Drucker said about the importance of management: "As the trees are rotten from the head and die, organizations are suffering from degradation and destruction when the managing director of that organization can't manage it." This statement applies to leaders of megacities. The importance of management and the strategic mind behind innovated leading initiatives become the essence of leadership as innovative management strategies enable managers to learn from previous management activities, and respond quickly and creatively to new challenges and risks.

All the approaches presented in this volume deal explicitly or implicitly with the question: How do we apply and deliver concrete strategies to confront the vulnerabilities of megacities to climatic events affecting public health? It also addresses the challenges of attaining sustainable development, capacity building, proactive and reactive strategic planning, and improving and maintaining the megacity's infrastructure and services to sustain resilience. Professor Roger Martin asserted in the *Harvard Business Review* (2014): "Strategy requires making choices about an uncertain future." A well-rounded strategy can predict the future of any organization accurately enough to choose a clear strategic direction for it, by applying a set of powerful analytic tools. Presently, climate change creates an uncertain future for megacities; approaching the strategic plan underestimating uncertainty can lead to a lack of strategic vision in concrete terms to advance the goals and objectives of the strategy making the megacity vulnerable against threats while also overlooking opportunities that higher levels of uncertainty provide. Another danger lies at the other extreme: "If managers can't find a strategy that works under traditional analysis, they may abandon the analytical rigor of their planning process altogether and base their decisions on gut instinct" (Courtney et al., 2000).

In planning a new strategic order, a "new way of analytical thinking and leading addressing world's issues that affect global communities, present and future generations," leaders and leading organizations must recraft the existing policies and challenge the status quo. Harvard Professor Clayton Christensen said in his book *The Innovator's Dilemma* (1997), which provides a set of rules for capitalizing on the phenomenon of disruptive innovation, "capturing new markets embracing new technologies and adopting new business models" is key to growth and innovation. It can be understood as unleashing the potential contribution of new strategic models for growth, progress, and resilience. Innovation, strategic flexibility, and sustainable capacity can help megacities' mayors address economic crises affecting growth and environmental challenges that continue to mount, as governments around the world increasingly seek more innovative ways to promote economic activity and enhance sustainability. New technological advancements and computer climate modeling, contribution from information and communication technologies to climate change mitigation and adaptation, and risk management frameworks that can tailor new ways of strategizing business models and allay risks are all under the same umbrella when embracing innovation and disruptive influences into a new realm.

A comprehensive strategic planning approach examines how successful models can be scaled up and accelerated through appropriate policy action and position leaders to understand that, without evolution and improvement, nothing survives. Climate change risks and its implications to humanity will drive "disruptive innovations" to confront these new challenges, adapt to a new reality, and build capacity for endurance. Any strategies adopted (or not) by world leaders can advance or inhibit the transition to a detailed approach to sustainable management and strategic planning. By all accounts, the social values, the urgency that shapes the implementation and development of the plan, and the sociocultural, socioeconomic barriers that impede the strategies to be successfully implemented will influence the decision-making process. "Motivation is the catalyzing ingredient for every successful innovation. The same is true for learning" (Christensen et al., 2011). Motivation lies along the need for change. Therefore, leadership competencies for disruptive innovation transform societies and changes humanity in all sorts of ways with profound impacts. Innovation implies substantial positive change, although all changes are not automatically innovation (Hoffman et al., 2011).

Innovative strategies and adaptive thinking go hand in hand with facing disaster management and severe climatic events. Transnational or local forms of action are required if a city is to translate its mission and values into actionable and measurable initiatives and objectives.

Capacity for change must be rooted in governance as megacities confront rapidly changing, often unpredictable environments. The need for leadership action to tackle the risks caused by climate change requires strategic adaptive frameworks that understand the likelihood and consequences of such impacts. Adaptive capacity means the ability "of a system (e.g., ecological or human social) to adapt if the environment where the system exists is changing" (World Water Assessment Programme, 2012). Heads of government, mayors, risk managers, executives, policy makers, environmental leaders, nation and community health leaders are all confronting a daunting unknown future without the proper means to address risk, defining the scope of risks to be tracked, and documenting and improving the process. These actions support a concise strategy and avoid having the analysis becoming stalled and not reflective of potential threats.

It is here that the strategic adaptive cognition (SAC) lays out the concept of "creating a strategic mindset that takes a broad, long-range approach to problem-solving and decision-making." Because the risks are broad and global, planning ahead, as well as understanding the current situation through objective analysis and pragmatic steps, is required ultimately to accomplish the strategic goals. It is important to consider a strategy of understanding and acceptance of different approaches to change while enhancing willingness and ability to change. SAC also helps to identify the impact of the leaders' decisions on the differing configurations of the megacity, which is the structural aspect of this complex institution. John R. Wells defines strategic intelligence as the capacity to adapt to changing circumstances, as opposed to blindly continuing on a path when all the signals in your competitive environment suggest you need to change course (Wells, 2012).

SAC's concept originates from problem-solving approaches and novel strategies that are often best learned analytically and experientially. There are seven steps:

1. Systematic thinking—Enhancing the ability of a person or organization to use learning effectively is crucial to comprehend the implications of any critical situation. Megacities can learn from one another's failures and successes by continually expanding their capacity to create the results they need to accomplish. When new and expansive patterns of thinking are nurtured, where collective aspiration is set free, and where people are continually learning to see the whole together, goals become tangible and results will be materialized.

2. Strategic approach to management and policy—A forward-thinking strategy that addresses the three responsibilities of an organization: social, environmental, and financial. It needs the proper systems implementation to pursue sustainability and evaluate the impacts of sustainability on social, environmental, and financial performance along with the trade-offs that ultimately must be made, putting emphasis on the present, considering future outcomes and long-term strategies to cope with climate change impacts.

3. Adaptive and flexible behavior—Many animal species employ behavioral flexibility as an adaptive response to changing environments. This flexibility lies at one end of a continuum of plastic responses that includes developmental plasticity in individual

physiology and anatomy, and genetic responses to selection over generations (Dukas, 1998; Pigliucci, 2001; West-Eberhard, 2003). Of these forms of response, behavioral changes generally occur most quickly and therefore are best suited for rapid responses to changes in the external environment. As the impacts of climate change put societies and populations in a changing environment that worsens existing risks and increase vulnerabilities, adaptation of leaders to environmental change affecting their organizations becomes a fundamental asset for understanding such threats and assimilating the necessary knowledge and information to find strategies and solutions to those problems. Strategic flexibility goes along with behavioral flexibility in adapting to external and internal changes and challenges that threaten the existing structures and systems that cannot deal with the new risks.

4. Strategic cognition—Learning from other leaders, strategies, and experiences and subsequently applying them to evolve your strategy for improvement. In the case of climate change impacts on megacities, past experiences provide the path to a particular course of evidence and then takes that into account to plan, do, check, and act to address climate change risks. This action requires being people-centric. Leaders need to have the ability to accomplish goals and objectives by operating in cross-cultural contexts with interpersonal skills, and adaptive thinking.

5. Critical thinking and learning—Assimilation and understanding information can be biased, distorted, partial, uninformed, or downright prejudiced. However, the quality of our lives and of what we produce, make, or build depends precisely on the quality of our thought. For leaders, fostering critical learning and thinking habits becomes crucial to being prepared to confront important issues, much more so than to simply learn facts. "Habits of Mind are the characteristics of what intelligent people do when they are confronted with problems, the resolutions of which are not immediately apparent" (Costa and Kallick, 2008).

6. Experiential training—Develop behavioral skills and objective abilities. "Learning by doing" involves the interaction of participants with the goal of assimilating and understanding specific situations and applying them to their own reality, unlike the informational training methods that are more one-sided. Here the major focus is not just the mere transfer of facts and figures but also the development of skills in the participants, which may or may not be the case in informational training.

7. Shared value and logical incrementalism—A change of mindset that requires the thinking of growth and the improvement of societal

conditions. Shared value applies to businesses around the world, and it can easily relate to megacities creating measurable benefits by identifying and addressing social problems that intersect with their organization implementing sustainability, social responsibility, and sustainable development. Shared value, writes Initiative for a Competitive Inner City founder and Harvard Business School Professor Michael Porter, is defined as "policies and operating practices that enhance the competitiveness of a company while simultaneously advancing the economic and social conditions in the communities in which it operates. ... Shared value is not social responsibility, philanthropy or even sustainability, but a new way to achieve economic success." When this success is inherent in a megacity's system, sustainable development rises and poverty eradication can become materialized (although, not all economic progress is equal to poverty eradication). Logical incrementalism goes along with shared value strategies as it states that strategies do not come into existence based on a one-time decision, but rather, it exists through making small decisions that are evaluated periodically. Such decisions are not made randomly but logically through experimentation and learning. Embracing their interdependencies within communities and strategically including community impact in a megacity's strategy can produce measurable advantages and leverage private development money. It is important to note that the work of city leaders is primarily made up of mediation and collaboration and has the unique ability to leverage both public and private money to create shared value (Initiative for a Competitive Inner City, 2011).

A vital tool for leaders is the learning to use "what you know." Cognitive learning requires repletion, summarizing meaning, guessing meaning from context, and using imagery for memorization. Recognizing problems and making changes to correct them often presents substantial challenges (Suharno, 2010). Integrating old knowledge with new knowledge is the path for a successful plan. The recognition of problems at an early stage is highly important as such problems will be magnified over time and more damage will occur to the organizations; consequently the difficulty to solve them will be greater. This can be exemplified by the "inverse square law," which states that a specified physical quantity or intensity is inversely proportional to the square of the distance from the source of that physical quantity (Appalachian State University, Department of Physics and Astronomy, 2015). Applying this concept to businesses and/or organizations leads to the understanding that an important issue that is not resolved promptly will multiply and become difficult to control and/or fix in the future. Therefore, a focus to resolving issues, determining and addressing risks of the megacity will need the capability to enact major strategic changes to resolve problems in a timely fashion; otherwise the cost of delay will exacerbate those risks. This approach requires strategic flexibility, which comprises three elements: maintaining attention, completing an assessment, and taking action (Shimizu and Hitt, 2004).

## Game Theory, Externalities, and the Strategic Thinker

Game theory is "a mathematical technique developed to study choice under conditions of strategic interaction" (Browning and Zupan, 1998). Maskin (2008) divided game-theoretic solution concepts into those that are noncooperative—where the basic unit of analysis is the individual player—and those that are cooperative, where the focus is on coalitions of players. Climate change presents many opportunities for using game theory tools to develop strategic plans for mitigating $CO_2$ emissions or minimizing its impacts. For example, integrating environmental conditions, such as air pollution, and other factors that require a negotiation process are appropriate targets for game theory applications. The representation of scientific knowledge regarding climate change and its impacts has an important significance for assessing political programs, policy reform, and evaluating the responsibility of governments and the public in addressing those risks. Because game theory deals with conflict, it can provide useful tools to strategize when considering debates and arguments when dealing with a set of collaborative initiatives or playing solo in initiating specific climate change schemes.

The principles of game theory can be applied to mitigation and adaptation programs because they are simple representations of multiple situations that climate risks present to cities around the world. For example, city mayors can lead the way by assembling a framework for adapting to climate change; or they isolate themselves and take no action. In the first case, leaders will confront different scenarios that weigh the greater public good between equal and unequal size groups. We can universally understand the need for an adaptation strategy to diminish impacts of climatic events. For the common good, developing and implementing a framework that meets those needs is crucial and important. At times, however, it is optimal for some leaders to let someone else take the initiative and absorb the costs and risks associated with implementing a strategy, allowing one city's operations team to learn from the experience of another city. In economics, this is known as "externalities" (Levine, 2015).

Externalities transpire when one city's actions affect another city's safety, incrementing the risks associated with no relevant strategic implementation. Hence, the initial investments to implement the plan as well as the gains (e.g., return on investment, reducing health impacts) are not reflected appropriately because one city's unique situation cannot translate equally into another city's "reality" (Levine, 2015). Obviously, we need to consider the geographic, environmental, social, and economic situation of the city, as these factors influence the outcome of any strategic management plan.

## Externalities and Decision Making

A positive externality is a benefit that is enjoyed by a third-party as a result of an economic transaction. Third parties include any individual, organization, property owner, or resource that is indirectly affected (Kaczorowska-Ireland, 2015). Positive externalities take place when a city benefits from successfully implementing an adaptation, mitigation, and risk management plan, which other cities then follow. A negative externality occurs when a city's lack of leadership (and initiative to develop a strategy) adversely affects another city or cities, and leads to a reactive after-the-event approach. These externalities pose a serious problem because of their potential for adverse consequences (Levine, 2015). A changing climate presents several choices for city leaders (the players in game theory) to consider in dealing with risks. In modeling activities (both proactive and reactive) for changing temperatures, it is important for the parties involved to know the characteristics of the events, actions, and behaviors. The certainty of the game's payoff is not established. However, actors' expectations must be considered as well as their possible reactions to one another or to the probability of certain external events. Particularly in the application of climate change, the uncertainty is accumulative. Using game theory terms, uncertainty means there is no saddle point in the game (Kutasi, 2010).

## The Uncertainty of Risk Aversion

Owen (2008) refers to uncertainty as the dominant characteristic in climate change. Risk management deals with the identification, assessment, and prioritization of risks, and lays down strategies to cope with uncertainties. Ellsberg (1961) conclusively defined uncertainty as an event that is uncertain or ambiguous because it has an unknown probability. Climate change poses an indistinct unknown outcome with projections of a future distance that can be challenging to quantify and estimate. With a mechanism design in place, a risk management framework benefits when faced with two or more strategic plans for climate change with a similar expected outcome, taking into account the different *risks*. Therefore, the strategy chosen will be the one with the lower risk.

## Mechanism Design Approach to Environmental and Health Risks

Professional economists have started addressing the environmental aspects of economic actions. Environmental issues influence and are influenced by aggregate economic actions along many dimensions (Razmi, 2016). The theory of mechanism design was created by Eric Maskin, Leonid Hurwicz, and Roger Myerson, who shared the 2007 Nobel Prize in Economics (Börgers et al., 2015). According to Baliga and Maskin (2003), because environmental issues often entail nonexcludable externalities, the theory of mechanism design (sometimes called "implementation theory") is particularly pertinent to the economics of the environment.

Mechanism design holds promise for more sustainable forms of economic processes. It is suitable for evaluating impacts on health and other issues. While designing the strategy for a megacity, planners can align the planning within the proposed risk management framework with a mechanism design theory approach to lay down a strategic plan so that the outcome is more like what is desired (Mehaffy, 2008). This approach has a few frames that can correlate to risk management strategies for continual improvement, and its application can go a considerable way to developing and implementing sustainable management and strategic planning, particularly in megacities where leaders face numerous hurdles to proposing economic, environmental, and social change as well as policy reforms to achieve sustainable development and sustainability. Many environmental economists have written about climate change and macroeconomics crossing paths between the paradigms of evolutionary economics and environmental economics dealing with decision making of an economy as a whole, rather than individual markets. The question arises, however, about what happens if the strategic rules change. Can regulators set up a model that emphasizes not only the benefits (in this case adaptation and mitigation strategies to deter climate change effects), but also the consequences, because mechanism design seeks to changing the rules of the economic game to shift the optimal outcome? (Mehaffy, 2008).

As a result, strategizing with a mechanism design approach in mind, whose equilibrium corresponds to an optimal outcome, can lead policy makers, mayors, and other decision makers to substantial cooperation based on conditional commitments.

## The Economics of the Environment: Climate Health Risks and Mechanism Design Strategies

Can cities manage climate change risks without government intervention? Is a national or international mandate needed to find cooperation and synergies to address such risks? Is COP21, and similar negotiation strategies on an international level, the only path to a comprehensive and effective framework for solutions to a common issue and goal? Megacities, because of their economic power, network of infrastructure, and mayors' influence in policy decisions in the local and global arena, can identify individual actions and cooperative initiatives toward managing these risks. However, the argument presented next specifies that a coercive authority must be called to demand a process for determining those actions toward a common goal.

Climate change and its implication for public health, environmental degradation, and risks often involve externalities that cannot be excluded; the theory of mechanism design is particularly pertinent to the economics of the environment. Baliga and Maskin (2003) argued that a pure public good, which, once created, will be enjoyed by everybody, constitutes the classic example of a nonexcludable externality. They extended the thought by describing what goes wrong with nonexcludable externalities, such as pollution. As an example, despite describing air pollution as a pure public evil and the outcome of any strategy to deter or minimize its detrimental effects on public health as a public good, many municipalities emit pollution affecting the overall health of its residents. Considering the different economic factors and incentives a city might have, a community might require a challenging and costly pollution emission reduction plan because it entails curtailing or modifying the community's normal activities. From an economic perspective, if the city acts alone, it will bear all the costs associated with the plan implementation, although it can share, with certain limitations, the benefits of the reduction plan. The ideal outcome would be to reach *Pareto optimality*. The outcome of a game is Pareto optimal if there is no other outcome that makes every player at least as well off and at least one player strictly better off. This will achieve a harmonious balance among players and will reduce, in this case, pollutants in the atmosphere on a larger percentage (considering most players are actively reducing emissions) than one player or a small number of players actively pursuing this reduction. However, a problem can arise when one city decides not to participate in the reduction plan. As stated by Baliga and Maskin (2003), by staying out, the city can enjoy the full benefits of the reduction plan (this is where the nonexcludibility assumption is crucial) without incurring any of the cost. This result assumes that the agreed reduction will be somewhat smaller than if that city had participated (since the benefits are being shared among only $N - 1$ rather than $N$ participants). However, this difference is likely to be small relative to the considerable saving to the single player from not bearing any reduction costs. Thus, it pays the city in question *to free-ride* on the others' agreement. A problem arises because this is true for *every* city in the reduction plan, so it is possible there will be no pollution-reduction agreement at all. These actions could potentially lead to all cities participating not implementing a reduction plan and could also lead to actions taken on an individual basis that will detrimentally affect the air pollution strategy, jeopardizing its efficiency. Analyzing this situation, one can fairly conclude that an external influential entity or coercive authority is needed to determine a method, called *mechanism* (or game form) to achieve the desired outcome (Baliga and Maskin, 2003).

### General Theory of Strategic Adaptive Cognition: Mechanism Design, Microeconomics, and Strategic Planning

Developing a suitable mechanism can create complications and challenges for implementing the desired strategy with specific outcomes. Several factors need to be considered when developing a mechanism design approach such as relevant information and other assets held by the participants in order to optimize their benefit that different players enjoy from pollution reduction. The efficiency of applying a mechanism design sits in the ability of designing the rules of the game so that the players are motivated to report their preferences openly when a (socially) desirable outcome is chosen. Megacities facing climate change impacts and its implications on health and the environment generally have different preferences of addressing risks where their budgets, allocation of resources, infrastructure system, and political and social environment dictate or influence their ability to be proactive and proceed efficiently. Moreover, the multifaceted situation they face can bring different players not being aligned with the core strategy. This is because developed and developing countries' cities are faced with a diverse pool of challenges that mirrors their geographical location and population vulnerabilities, political and socioeconomic environments, and other factors, influencing how the strategy will be implemented, if implemented at all, and how leaders behave and act upon those challenges. Strategic planning and strategic cognition play an important role in identifying efficient paths toward a positive outcome. One thing that we know from psychologists is that actual people are not as fully rational as theorists will like them to be. In reality people make mistakes in decision making that is important to take into account when designing mechanisms. In mechanisms, companies, institutions, or people have to take some actions and you want the mechanism to still work reasonably well even if people or institutions make mistakes in the actions they take (Maskin, 2013).

Establishing a link between climate change, public health risks, and a strategy that encompass a mechanism design (implementation theory) to ensure a Pareto-efficient outcome is a challenging but viable action. Mechanism design presents a different outlook where it starts identifying the outcome that the institutions/players/agents would like to have and goes backward to figure out which institutions will generate those desired outcomes and as Professor Maskin referred "it is economics in reverse."

Now, the benefits of mechanism design and its potential to be applicable in the study of the environment and the solution of environmental problems is still in an early stage. The understanding of the consequences of climate change impacts in megacities' dwellers and the ways decision makers are dealing with risks posed by those impacts implies a far more radical rethinking of structure and strategic planning than has been realized to date. Strategic adaptive cognition identifies and rationalizes empirical pattern recognition to help challenge existing strategies of decision making and guide the foundations of new theories of choice. Megacities (the system) are a network of systems (subsystems) within a large system of interconnected networks. These subsystems comprise a network that operates collectively in society for an (ideally) efficient functionality. Using a strategic adaptive cognition plan and applying the concepts of mechanism design, players interact with knowledge of one another under a set of common understandings. They are aware of what needs to be accomplished (for example, pollution reduction), the relationships among the players, and the rules imposed by their current political and socioeconomic situation. These relationships affect the mechanism design outcome and have an affect on the implications of our analysis for modeling collusion between multiple agents interacting with the same principal.

### Thinking Like a Mechanism Designer with Strategic Adaptive Cognition in Mind

A megacity's leadership team can think like a mechanism designer and can make several players, interest groups, and public and private sector entities interact with each other. In this way, the city can find a path to solve specific problems, find common ground, and become integral to organizational improvements and policy, thus providing a coherent framework for analyzing this great variety of institutions, or "allocation mechanisms," with a focus on the problems associated with incentives and private information (Prize Committee of the Royal Swedish Academy of Sciences, 2007).

Thinking like a mechanism designer requires defining the game and setting it up. In mathematical terms that means optimizing the allocation of resources when the optimization parameters needed to determine allocation are privately held by the agents who will consume the resources (Vohra, 2012). The players involved will do what is more beneficial for them, taking into account that the players in a major city (public and private sector, policy makers, environmentalists, mayors, community boards) will behave as they want, not necessarily as we expect them to. As the example of the air pollution shows, the cohesive authority (the one who imposes the rules) cannot expect the players to be truthful, especially because it is hard to determine what percentage of pollution reduction can be estimated or what action will be taken to reduce any environmental degradation that affects public health within the community. When designing mechanisms, people must be expected to behave strategically.

The central elements of the theory are as follows:

- Designing a mechanism for any interaction between people/institutions/organizations.
- Designing institutions that satisfy certain objectives
- Choosing the right rules of their mechanisms

- Inducing efficient trade of the good, so that successful trade occurs whenever the buyer's valuation exceeds that of the seller
- Defining the objectives (Jackson, 2003)

The core principle of the theory is to use a strategy in a framework of actions leading to the desired outcome. Mechanism design has adaptive synergies that can tailor a risk management approach to the effects of climate change. To achieve systematic and continuous improvement requires analyzing and applying a framework using mechanism design (a risk manager as a mechanism designer) to provide a mathematical abstraction to cope with different problems. This action should include combining the existing ad-hoc schemes under a single umbrella leading to solutions that have a systematic continuity for improvement. In essence, a mechanism design-risk management framework is a way of implementing new ideas in a controlled way. Interactions between players are key to a successful outcome. What comes out of those interactions between players is the goal. Defining the goal is crucial. Then, when designing the system of interaction, the incentives that this will create are the next step. Two considerations arise from this methodology:

1. The reaction of the players toward the rules the mechanism designer presents
2. Consequently, considering the reactions of players, what the outcome will be

The SAC framework (micro, meso, macro levels) is a mechanism to help accomplish a particular *cognitive* goal, articulate the goal, and understand and accept different approaches to reach an outcome. Understanding and acceptance complement mechanism design and encourage a willingness and ability to change. SAC also helps to identify the impact of the leaders' decisions on different game configurations, which is the structural aspect of the complex process that players in the game can encounter during negotiations, planning a strategy, and other game moves. These steps are critical when all signals of the environment are leading to change. Therefore, the SAC concept originates from problem-solving approaches and novel strategies that are often best learned analytically and experientially. A mechanism design problem has three key inputs:

1. A collective decision problem, such as the allocation of work in a team, the allocation of spectrum for mobile telephony, or funding for public schools
2. A measure of quality to evaluate any candidate solution, for example, efficiency, profits, or social welfare
3. A description of the resources—informational or otherwise—held by the participants (Legros and Cantillon, 2007)

## Strategy and Mechanism Design

The challenge for mechanism designers and its implementers is that scenarios vary according to the vulnerabilities faced by countries or cities, communities, and their capacity building as well as their economic and social stability to cope with such risks and the involvement of stakeholders (public and private) that are the engine of megacities. In mechanism design all players "agree" to a common goal. Climate change negotiations are complex simply because all players involved have different needs, interests, and agendas. Now, with a strategic adaptive cognition planning, mechanism design can benefit using the micro, meso, and macro levels of strategic implementation because these three levels consider the microeconomic context and environment where the nations/cities/communities operate. COP21 negotiations in Paris in December 2015 made important progress toward international agreements where all players can find common ground to reduce their emissions. In cities where the environmental, social, and economic situations have a microeconomic frame (use of microeconomic approach to target resource management issues), the international agreement has certain limitations to be implemented as the interests of local, city-level stakeholders play an important and influential role in decision making.

Bergemann and Morris (2012) provided an insight by asking "whether it is possible to explicitly model the robustness considerations in such a way that stronger solution concepts and/or simpler mechanisms emerge endogenously." If the optimal solution to the problem the players are trying to solve is too complicated or too sensitive (climate change, environmental pollution, health risks) to be used in practice, it is seemingly because the original description of the problem was itself flawed. Mechanism design dominant strategy is based on the systematic look at the design of institutions. These organizations/groups/players determine and influence the outcome of the decisions made. The main focus of mechanism design is on the design of institutions that satisfy certain objectives, assuming that the individuals interacting through the organizations will act strategically and may hold private information that is relevant to the decision at hand (Jackson, 2003).

Game theory is a strategic interaction between two or more players. Furthermore, it is a mathematical theory and, as such, provides many rigorous models of interaction and theorems to specify which outcomes are predicted by any given model. Game theory and subsequently mechanism design can provide the modeling frameworks to assist in efficiently reaching the desired goal. However, the type of strategic reasoning underlying climate change and environmental risks does not directly leverage deep mathematics. Instead, this reasoning is a pathway that leads to actions where all players reach a common goal in different ways.

Not having a choice among players can be also considered. The traditional approach to strategy requires precise predictions (Strategy Under

Uncertainty, 2000). Hence, a strategic adaptive cognitive planning assists to "framework" your analytical thinking and materializes it to display concrete actions that lead to the implementation of the strategic plan without having a "choice" for all players involved.

Mechanism design involves interactions taking place in closed, controlled environments, where the rules of the game are specified exactly. Strategic adaptive cognition supports the understanding of the game involving objective and pragmatic analysis that ultimately accomplishes the strategic objectives. If the game is played in closed, controlled environments, the players, as well as the cohesive authority (the one who sets up the game's rules), must consider a strategy of understanding and acceptance of different approaches to change, enhancing willingness and ability to change. Understanding that humans are unpredictable, the game theory and mechanism design theory works better with SAC for global climate change that affect people and the environment. What is unique about this SAC perspective on risk and disaster is the holistic perspective, which points to reduce vulnerability by addressing identified gaps in the strategic approach and encouraging involvement of and collaboration among all sectors of societies and stakeholders to achieve effective disaster and health risks reduction. This way, when the signals in your competitive environment suggest you need to change course and adapt to changing circumstances, rather than blindly continuing on a path that is "framed and rigid," the use of SAC planning and its methods will identify the weaknesses and strengths of the mechanism design framework. Then, in a real-life situation, it is easier to overcome obstacles and challenges in decision making.

Consequently, because the players are aware of these strategic interactions, they will have a complex task determining their best response and course of action. In supporting a mechanism design framework, SAC has the advantage of considering and acting on the strategic behavior between players that will influence each player's decision-making strategy addressing the authority in charge of the game's rules, and with each other.

The dominant strategy equilibrium (regardless of what the other individual does, each individual receives a higher pay-off from defecting than from cooperating) has been played by climate change, environmental leaders more often and frequently. The dominant strategy to address climate change presented by each organization, regulator, nongovernmental organizations (NGOs), and/or interest agents presenting their own strategy faces challenges for governments to rapidly reduce emissions. Unfortunately, existing strategies in the megacities does not capture very long-term structural changes, and businesses and civil society are likely to expect policy shocks. Some countries and cities will experience policies that make carbon-intensive activities more expensive. Therefore, businesses and civil society will need to be able to respond to these shocks. Although climate change cannot be addressed with one single strategy, it is crucial to unite world leaders under frameworks for local and national governments to drive down carbon emissions to avoid extreme climate change and prepare for its impacts is crucial. Megacities need to agree on a framework on the basis of equity and in accordance with their common but differentiated responsibilities and respective competence to take suitable actions.

The most significant international climate change agreement—COP21 in Paris under the United Nations Framework Convention on Climate Change—represents cohesive authority laying down the rules and each party (player) adapting to the main goal according to their realistic expectation and common reality of events, factored by economic and social outcomes. The dilemma comes when all players within the megacities that have sufficient influence to reform policy and implement climate-resilient strategies have no incentive to changing their strategy given every other player's strategy.

Strategy involves understanding the mechanism that describes the objectives, concepts, tasks, and capabilities necessary in the near or long term to adapt and improve the organization's proven capabilities to meet a challenge. The understanding of the megacity's approach to address the challenges of climate change, and the efforts of decision makers to accomplish their desired goals, is vital to develop a strategy that will not lead to a "war of attrition." The players cannot engage in a prolonged period of conflict in which each decision maker tries to gradually wear out the other by a series of small-scale actions. Game theory suggests that the likelihoods and targets of a future outcome can be modeled by understanding the operational and behavioral characteristics of the organization. It involves the concept that players are rational and make choices based on information and rules. Is game theory a mechanism to get inside the minds of the players involved (Purpura, 2007)? Game theory and mechanism design are compatible and interconnected to risk management frameworks. Many organizations are increasingly utilizing game theory to help them make high risk–high reward strategic decisions. That applies to a highly competitive market within a global context and it can also apply within the framework of climate change negotiations. Where game theory can generate the ideal strategic choice in a variety of different situations, a set of strategies accommodating well-defined mathematical scenarios must be available to decision makers. Also, the payoff for each combination of strategies must be known. Typically, multiple strategy games are played to model different competitors, various payoffs, and potential strategies. Once the desired outcome and target relationships are set, the strategic decision can be modeled. The consideration of potential choices and payoff of others is essential when deciding a strategy, and must take into account that other players will consider your strategy as well. This understanding—quantified through payoff calculations—enables the parties involved to formulate their optimal strategy (Erhun and Keskinocak, 2003). Consequently, mechanism design is optimal for strategic situations where competitive or individual behaviors can be modeled. Mechanism design is ideal to clarify strategists' thinking and planning. A properly constructed game with a defined mechanism can reduce risk, improve the plan–do–check–act process in risk management frameworks around decision-making, and maximize strategic utility.

Many megacities began to seize on potential solutions for complex global problems such as climate change effects. Facing a particular issue expanded the environmental impact capacity of each megacity and ultimately shifted attention to how they collectively address an issue that might be global in scale but affect each megacity differently. Obstacles for agreements and aligning a strategy arises because it requires complex understanding of cross-cutting issues and therefore decision makers need to efficiently coordinate the efforts between the different levels of departments within the megacity's functionality, information management, climate risks research, cooperation and integrated management approaches, and so on. When information differs across different players, the process to accomplish the objective becomes more complex, and the psychological elements in play need to be mastered to successfully implement optimal plans. Achieving alignment during the implementation phase doesn't warrant that the strategy won't fall apart. A mechanism design is very well equipped because it balances the costs and benefits between the involved players, and this balance can be reflected in the design of the cooperation mechanisms. As an example, Dasgupta, Hammond, and Maskin (1980) considered an economic environment in the context of a simple model of pollution control consisting of $n$ firms $(i, j = 1, \ldots , n)$. Firm $i$ faces a cost function $C\,(x_i \theta_i)$, where $x_i$ is the firm's pollution emission level $\left(x_i \in R_+^1\right)$ and $\theta_i$ is a parameter (possibly a vector) known to the firm but not to the regulator. Let $\Theta_i$ be the set of possible values of $\theta_i$, and take $\underline{\theta} = (\theta_1, \ldots , \theta_i, \ldots , \theta_n)$ and $\underline{\theta}_{-i} = (\theta_1, \ldots , \theta_{i-1}, \theta_{i+1}, \ldots , \theta_n)$.

Furthermore, let $\underline{x} = (x_1, \ldots , x_i, \ldots , x_n)$, and $\underline{x}_{-i} = (x_1, \ldots , x_{i-1}, x_{i+1}, \ldots x_n)$.

In what follows we shall on occasion write $\underline{\theta} = (\theta_i, \underline{\theta}_{-i})$ and $\underline{x} = (x_i, \underline{x}_{-i})$. Suppose that the social damage caused by the vector of pollution levels $\underline{x}$ is $D(\underline{x})$. We shall assume that the regulator can monitor $\underline{x}$ without cost, and that if he knows the true value of $\theta$ he would wish to choose $\underline{x}$ to minimize the sum of damage and total costs $D(\underline{x}) + \sum_i^n {}_{=1} C(x_i, \theta_i)$. Presume that a unique minimum exists for every possible value of $\underline{\theta}$. Let $x_i^*(\underline{\theta})$ $(i = 1, \ldots , n)$ be the optimal level of pollution for firm $i$. It is the full optimum; it bears emphasis that $x_i^*$ is a function of the vector $\underline{\theta}$. But by hypothesis, the regulator does not know, a priori (based on theoretical deduction rather than empirical observation), the true value of $\underline{\theta}$. One is, therefore, encouraged to search for optimal tax-subsidy schemes in the face of this initial lack of information. In particular, it is important to know whether the full optimum can be attained despite this lack (Dasgupta et al., 1980).

It is also important to consider accountability when group-level goals are missed. Edward H. Clarke (1971) and Theodore Groves (1973) are frequently cited as pioneering an economics literature on mechanisms that, at least in principle, could encourage people to truthfully reveal their willingness to pay for public goods (Falaschetti, 2013).

Mechanism design is optimization without information and concisely allocates resources efficiently. All the information is with rational players, while the decision maker is uninformed. Coming from this analysis the important factors to be considered when designing a mechanism with optimal results is the understanding of social costs and risk implications. The strategic approach is addressing an environment of multiagent settings, and the optimal outcome needs to be based on the preferences of multiple self-interested agents, who will not necessarily report their preferences truthfully if it is not in their best interest to do so (Conitzer and Sandholm, 2007). Figure 22.1 represents a strategic approach to designing a mechanism for the megacity's mechanism designers where players do not accurately report their preferences. This is a strong case when addressing environmental risks and climatic events as the most relevant/accurate climate science is not always immediately available because scientific conclusions are always revisable if warranted by the evidence. The flowchart in Figure 22.1 represents the process of mechanism design models that can be scaled up and accelerated through appropriate policy action and strategic planning.

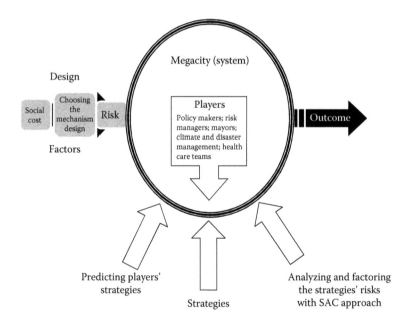

**FIGURE 22.1**
Megacity's mechanism design with strategic adaptive cognition (SAC) framework. (From Marolla, C., Megacity's mechanism design with strategic adaptive cognition (SAC) framework, Scribd Inc. 2016.)

## Design Factors

While setting the strategic phase, the developmental process comes with the knowledge and experience taking into consideration the changing technology, social structures, monetary economic conditions, and so on. The mechanism designer as well as the players should reevaluate the parameters upon which they are basing their decisions. The utility functions and bidding strategies of heterogeneous secondary users must be defined. The model must represent the situation addressing the problems that are investigated (Zhong et al., 2014).

## Social Cost

Players must consider the expense to an entire society resulting from taking actions on a specific event, an activity, or a change in policy (U.S. Environmental Protection Agency, 2010). When assessing the overall impact of such actions in terms of social costs, decision makers in megacities can take socially responsible initiatives that should take into account direct impacts, environmental deterioration, and its cost expenses, as well as any indirect expenses or damages borne by others.

## Choosing and Setting the Mechanism Design

Mechanism design means considering important factors such as the level of uncertainty, likelihoods and consequences of risks, and the players involved in the process that need to be identified. The characteristics of the players as well as the information unknown need to be determined. The designer needs to know the players' utility functions, which are often infinitely dimensional, in order to choose the appropriate pricing and allocation rules, capabilities, the state of the problems that need to be addressed, and their beliefs (Chorppath and Alpcan, 2011).

In summary, a design setting consists of

1. The players involved
2. The potential types of each player, representing all relevant private information the agent has
3. The potential outcomes available
4. The players' preferences over each outcome for each type, possibly expressed as a utility function
5. The beliefs of each agent as a function of their type
6. A theory about the behavior of agents (Crawford, 1990)

## Risk

Allocate risk on an agreed basis between the public sector, the private sector, and the players involved in the negotiation. Identify risks that are addressed from climate change impacts (direct and indirect such as $CO_2$ emissions, air pollution, environmental degradation, etc.). Moreover, players need to identify infrastructure risks, uninsurable risks, and secure a robust balance among all parties in view of sharing risks and benefits. Benefits and risks associated with large energy projects have to be considered from an economic, environmental, engineering, and societal point of view. Decision makers can design mechanisms to help reduce climate change impact risks at less cost in addition to minimize risks by combining traditional and scientific management of the coastal ecosystem following a risk management framework to reduce the effects of natural and climate-change-induced disasters. The theory of mechanism design provides many different applications; for example, in deciding whether regulation, tort liability, insurance markets, or a mix will be most effective in managing specific risks to human health and safety (Cox Jr., 2002).

## Strategic Adaptive Cognition Approach

Analyze how individual preferences can be aggregated into the preferable outcome bearing in mind desirable social or collective decisions. Strategic adaptive cognitive planning follows mechanism design in considering one primary user and multiple secondary users coexisting with the primary user. The mechanism design framework seeks to execute equilibrium strategies, and the SAC approach toward planning is designed to alter the structure of the game to shift the equilibrium. The latter idea is the same concept that underlies contract theory in economics. When parties have an incentive to defect on a mutually agreed upon arrangement, binding contracts can change incentives such that the contractually compliant move is more advantageous than defection, because of the penalties for noncompliance. Contracts change payoffs and, hence, change the game structure (Gangestad and Simpson, 2007). SAC focuses on key issues instead of on the distracting information that can conceal clear thinking and a path to a successful strategic planning, and emphasizes the analysis of decision-making in situations where one individual's best action depends on the actions taken by other individuals.

## Risks and Outcome

A mechanism, or game form, is then an institutional structure that assigns an alternative (or a probability distribution over alternatives) to actions within this structure (Ghosh and Kleinberg, n.d.). The goal of optimizing a preferable outcome according to the threats associated with the strategy should be viewed as adjusting for risks to different individuals and addressing adverse-selection problems. Solving risks in an explicit optimizing framework magnifies the power of risk-adjustment policies. Precisely, conventional risk adjustment is never the optimal policy for a mechanism designer in a market with asymmetric information. Therefore, rational agents will apply bidding strategies that are at equilibrium, meaning (roughly) that no single agent can improve his/her outcome by modifying his strategy. The main goal of a mechanism designer to optimize the best possible outcome is to design a mechanism that achieves good performance at every equilibrium of bidding behavior (Lucier and Syrgkanis, 2015).

In synthesis, mechanism design is the subfield of microeconomics and game theory that emphasizes the process of how to implement good systemwide solutions (Yi and Cai, 2016). This strategic approach has found many applications in recent years and is relevant to the functioning of any modern-day economy, no matter the economic system being pursued. Mechanism design is able to measure awareness and proactively strengthen a risk management framework and create a guiding principle for incentives and accountability, enabling the alignment of interests and incentives of the megacities' leaders in improving the public health system addressing climate change risks. Because the behavior of players and/or participants can strongly diverge from the expected outcome when different situations are presented, a comprehensive step-by-step strategic process is highly recommended to support the mechanism in place. Systems frameworks, such as plan–do–check–act (PDCA) or the Deming cycle process and ISO International Standards, suggest ways to resolve any conflicts among discrepancies within the strategic approach taken. These frameworks are an effective way for carbon offsetting, for example, because it requires a large number of players to design, implement, and operate these types of projects that require stakeholders, authorities, policy and decision makers to coordinate adaptability, flexibility, and sustainability in today's global economy and human health risks. By considering mechanism design with the three levels of strategic adaptive cognition planning (micro, meso, and macro levels) the strategy contributes to a better understanding of managerial decision making regarding risks and it develops the underlying cognitive determinants of the position that megacity's leaders adopt when taking into consideration the likelihoods and consequences of climatic impacts. Mechanism design and strategic adaptive cognition planning offers a cognitive rationalization for leaders to push for radical change when addressing issues that are characterized by complex, conflicting yet interconnected global risks such as climate change.

## Conclusion

Individual actions cannot resolve collective problems. Therefore, a cooperative approach to respond to climate change is needed. Actions toward adaptation to and mitigation of climate change impacts on health are generally warranted by the risks this issue presents to megacities. Climate change is on the forefront of public attention, and the risks affecting entire populations are clearer now than ever. Although climate change has received a fair amount of consideration—particularly after several droughts, flooding, and hurricanes in the United States, and events around the globe in recent years—a comprehensive strategy to address climate change and public health is not in place. It should be of utmost priority to implement it, as health professionals have the responsibility to prevent health risks and provide proper treatment. Climate change provides opportunities to world leaders to develop and implement policies and pragmatic initiatives that particularly target the poor (and as a consequence benefit the whole population), as their vulnerabilities are exacerbated by climate change. Climate change impacts can create a global chain reaction of effects that would be devastating for the world community.

The identification risks and their repercussions on public health is an important step for developing suitable actions. Policies made to address climate change can be adopted more speedily and efficiently as the evidence of the effects on health become more palpable. The awareness of the health risks to humanity will lead to a comprehensive strategic plan for megacities and will build the case for policies and procedures that directly aim at the problem. The increasing unpredictability of severe weather events and the consequences of climate change, like higher temperatures and changing landscape, make systematic strategies more challenging in efficiently addressing risk. Rapid change requires flexibility and creativity, which are seldom characteristics associated with formalized planning (Cunningham and Harney, 2012).

Most cities view strategic planning as a static framework that is neither adaptive nor directive. Implementation of risk management frameworks with a strategic adaptive cognition planning approach builds resilient systems, communities, and institutions that are robust, adaptable, and have the capacity for rapid recovery. Michael Porter (2015) says that it "is strategically important to make change. … Any project or any program is just inevitably embedded in a strategy for the overall organization and ultimately if we don't understand that strategy we can't decide whether we have the right project or program. If we don't have clarity on the strategy we can't decide whether we got the right goals and specifications. … If you don't have a clear sense of the overall strategy it is very hard to connect a project or program to what's going on elsewhere in the organization in a way that ultimately delivers value."

Strategies to develop resilient societies and risk management frameworks are reciprocal and mutually connected. The risk management doctrine applies to a diverse net of entities that address risks in different types of environments but share a common goal of coping with hazards in an efficient way. Risk management strategies contribute to the achievement of resilience by identifying opportunities to build resilience into planning and resourcing in order to attain the goal of minimizing risk, while enabling the mitigation of consequences of any disasters that do occur (Risk Management Fundamentals, 2011).

Risk management strategies present a semistructure framework for action that allows simple and complex structures to operate without constraining. International standards are ideal to address risk because they require no reservations regarding the reuse of the standard and provide multiple implementations according to specific conditions, likelihoods, consequences, and level of risks. Assessing climate change impacts with a risk management strategy presents multiple scenarios; planning for different outcomes provides the basis to foresee alternative views of the future impacts addressing different patterns of the key variations of such risks. Developing a plan and sticking to it is not viable because the impact of climate events, and the complexity of our physical and social systems, requires a comprehensive strategic thinking and planning. Leaders  of megacities need to integrate synergies into existing institutional mechanisms, creating incentives for sustaining innovation. Megacities' residents need to appoint leaders to develop and implement priority cross-cutting initiatives into rapid adaptive and flexible actions. Leaders must learn quickly about what actually works and what needs to be done to improve risk processes and practices.

We are confronting a challenge that has not been faced before by human beings, and the world community cannot avoid the threats, as the consequences of inaction will be costly. We have an obligation to present and future generations to act accordingly to prevent the far-reaching effects of climate change, as we have the resources, determination, and leadership to address the severity of the imperilment. The success of deterring climate change and its impact on the health of the urban populations will depend on how rapidly and efficiently global adaptation and mitigation strategies are implemented. World leaders need to find a common ground of understanding this critical issue, as our survival depends upon the immediate actions of all parties involved.

The fundamental approach to the adverse risks posed by climate change to the world's welfare is management, which requires a comprehensive and all-inclusive *sustainable management* and *strategic planning approach*. The analysis, projections, processes for improvement, and objectives for leadership change that are presented in this volume will enable world leaders, scientists, policy makers, health care workers, academics, and planners to take bold and innovative approaches to the problems of climate change. These actions should prioritize protecting the ecosystem and natural resources, and promoting social and economic opportunities. This integral relationship provides the tools to critical planning closely related to sustainable development and economic, environmental, and social progress. The *strategic planning* presented provides a forward-thinking approach, confronts and tackles those issues, and sits at the core of setting up a concrete method and direction to achieve specific goals. Therefore, this book offers a series of highly important strategies and innovative approaches to the common concern, as well as a thoughtful description of climate change health risks and subsequent consequences of their impacts on people.

Regarding strategic approaches, the International Organization for Standardization (ISO) develops high-quality international standards. The standards do not seek to establish public policy, are not prescriptive, and bring substantial value with efficient and cost-effective tools in support of the implementation of public policies and as an element of good public governance. These frameworks can be voluntarily adopted, and support sustainable and equitable economic growth, which translate into sustainable development for megacities; promote innovation; and protect health, safety, and the environment. International standards provide much of the technical detail and safety requirements required to make a good policy work (ISO, 2011). Strategic adaptive cognition planning strengthens risk management frameworks by placing leaders in an advantageous position for developing and implementing policies and programs that are aligned to the strategy toward a sustainable and resilient megacity.

Strategic and adaptive cognition set the behaviors and conduct for cognitive strategies to carry on specific methods that leaders can use to solve problems, including all sorts of analysis, interpretation, and planning capabilities and reasoning, resulting in complex goal-directed actions for efficient performance and results. More important, the adaptive strategic aspect of SAC related to the behaviors and conduct of a person provide the ability to act in dynamic and shifting situations, which is critical for the survival of all living species, including humans.

The strategic mind behind the sustainable management and strategic planning developed in this book examines how the mitigation, adaptation, and disaster risk management approach affect the megacity system, and outlines the likelihood and consequences of impacts and the crucial initiatives to reduce risk and put forward an adaptive, yet flexible strategy to combat climate change. Strategic adaptive cognition and strategic adaptive cognition planning strengthen leaders' abilities to carry on the strategy and mobilize people, resources, and initiatives to achieve a common goal, and to measure and manage the plan implementation and actions to ensure the megacity's goals are met. It allows the strategic, operational, and scientific process to address and act toward climate risks to work without political meddling.

A radical paradigm shift in the way of thinking and acting is necessary to sway the course of climate change impacts on the urban population's health. Now is the time to implement proactive rather than reactive management of climate change risks, and for world leaders to claim the validity and significance of sustainability and to advocate for the well-being of the global community.

# Appendix: Climate Action in Megacities

Regardless of disaster management and emergency aid actions when confronting severe climatic events, megacities' efforts in response to rapid-onset disasters remain largely ad hoc and not accord to confront risks with the level of urgency and efficiency required, resulting generally in inappropriate and/or delayed actions. The megacities' challenge is to include deterministic scientific knowledge within a risk management framework of probability. Mayors, risk managers, scientists, policy makers, and members of the business sector and the community, in addition to the involvement of other relevant stakeholders are crucial to develop a tangible strategic planning to address risk (Table A.1).

## Climate Change Risk Assessment for Megacities

The climate change risk assessment form is not meant to be prescriptive but may be tailored to the individual situation (e.g., removing control measures which are not relevant and adding others which are) so long as the format is adhered to. This form may be used as a prior risk assessment, reviewing current ones in addition to supporting actions to manage risks when a climatic event occurs (Table A.2).

## Climate Change Health Population and Critical Services Interruption Risk Assessment Impact Example

Consider how many people will be affected and the impacts on vulnerable populations involved (e.g., elderly, children, poor). Additionally, consider if the megacity's system of response and adaptation is in place.

Table A.3 shows the adaptive capacity and risk management plan in place for managing risks of weather- and climate-related disasters.

## Obtaining the Likelihood Score

The likelihood score (1–5) can be obtained from Table A.4 for each hazard and then noted in the boxes provided.

**TABLE A.1**

Population, GDP, and GHG Emissions for the Megacities under Study

| Megacity | Country | Population (millions)[a] | GDP (US$bn)[b] | Total GHG (MtCO$_2$e) | Total GHG (tCO$_2$e/cap)[c] | GHG per GDP (ktCO$_2$e/US$bn) |
|---|---|---|---|---|---|---|
| Mumbai | India | 18.84 | 126 | 25 (est) | 1.3 (est) | 198 |
| New York | USA | 18.65 | 1133 | 196 | 10.5 | 173 |
| Los Angeles | USA | 12.22 | 639 | 159 | 13.0 | 249 |
| Lagos | Nigeria | 11.70 | 30 | 27 (est) | 2.3 (est) | 893 |
| Rio de Janeiro | Brazil | 11.62 | 141 | 24 | 2.1 | 173 |
| Beijing | China | 10.85 | 99 | 110 | 10.1 | 1107 |
| London | UK | 7.61 | 452 | 73 | 9.6 | 162 |

*Source:* World Bank, 2010, *Cities and Climate Change: An Urgent Agenda*, Washington, DC: World Bank.

[a] "The 2006 population figures are based on censuses carried out between 2000 and 2005 and adjusted to take account of average annual population changes." Available: www.citymayors.com.

[b] GDP figures are for cities and their surrounding urban areas for the year 2005 based on research conducted by PricewaterhouseCoopers. Available: www.citymayors.com.

[c] GHG per capita values are from the "City GHG Emissions per Capita" table (available: www.worldbank.org/urban). GHG per capita values presented in italics (est.) are national values, as city values are unavailable. The corresponding GHG emissions should be considered GHG indications, not specific city values. (World Bank, 2010, *Cities and Climate Change: An Urgent Agenda*, Washington, DC: World Bank.)

**TABLE A.2**

Researched Megacities Climate Change Vulnerabilities and Health Risks Impacts

| Megacity | Level of Confidence | Level of Likelihood | Level of Consequence | Climate Health-Related Risks Projected Impact |
|---|---|---|---|---|
| London | Medium–high confidence (>70%) | Likely to occur sometimes (occurred infrequently). | Major and widespread decline in services and quality of life within the community. Infrastructure and communication network severely damaged. Distribution of foods and other basic services paralyzed and can lead to a chaotic state. | Summer overheating potentially contributing to heat-related health problems. Premature deaths due to hotter summers are projected to increase and quickly overwhelmed public health and social care services. Increases in the frequency of flooding affecting people's homes and vulnerable groups (e.g., those affected by poverty, older people, people in poor health, and those with disabilities). Climate change implications affect public health, the continuity of health and social care services both within the National Health System and beyond, the resilience of local emergency services, and the most socially vulnerable (Walker, 2014). |
| Mumbai | Very high confidence (>90%) | Likely to occur many times (occurred frequently). It will be continuously experienced unless action is taken to change events. | Severe and widespread decline in services and quality of life within the community. Severe, semipermanent, and widespread loss of environmental amenity and likelihood of irrecoverable environmental damage. The region would be seen as very unattractive, moribund, and unable to support its community. | Death and injury due to flooding; reduced availability of fresh water due to saltwater intrusion. Contamination of water supply through pollutants from submerged waste dumps; change in the distribution of disease-spreading insects. Health effect on the nutrition due to a loss in agriculture land and changes in fish catch; and health impacts associated with population displacement (Patil and Deepa, 2007). |

*(Continued)*

**TABLE A.2 (CONTINUED)**

Researched Megacities Climate Change Vulnerabilities and Health Risks Impacts

| Megacity | Level of Confidence | Level of Likelihood | Level of Consequence | Climate Health-Related Risks Projected Impact |
|---|---|---|---|---|
| Lagos | Very high confidence (>90%) | Likely to occur many times (occurred frequently). It will be continuously experienced unless action is taken to change events. | Severe and widespread decline in services and quality of life within the community. Severe, semipermanent, and widespread loss of environmental amenity and likelihood of irrecoverable environmental damage. The region would be seen as very unattractive, moribund, and unable to support its community. | Temperature increase of 0.04°C per year from now until the 2046–2065 period, wetter climate that will result in annual rainfall, and longer rainy seasons and extreme weather and heat days. The expected impacts of these extreme events are loss of livelihoods, loss of land to the sea due to sea-level rise, loss of physical infrastructure, loss of ecosystems and biodiversity, pollution of surface and groundwater, increased risk of waterborne diseases, chronic respiration illness, corrosion, acid rain, and damage of crops. Growing evidence of relationship between air pollution and mortality, hospital admissions for respiratory or cardiovascular disease, and an associated increased risk of myocardial infarction in Lagos. Apart from the direct effects of the air pollution, the flooding consequence has a lot of influence on human health. Infectious water-borne diseases are caused by pathogenic microorganisms that are most commonly transmitted in contaminated freshwater. Flooding is the leading cause of water-related mortality and can result in vector associated problems such as the increase in mosquito populations, which translates into more mosquito-related diseases (Kazama, 2012). The effect of water pollution from the industries (Sujaul et al., 2013) and their health challenges are overwhelming. |

*(Continued)*

**TABLE A.2 (CONTINUED)**

Researched Megacities Climate Change Vulnerabilities and Health Risks Impacts

| Megacity | Level of Confidence | Level of Likelihood | Level of Consequence | Climate Health-Related Risks Projected Impact |
|---|---|---|---|---|
| New York City | Medium–high confidence (>70%) | Likely to occur; sometimes (occurred infrequently). The activity or event expected to occur 50%–99% of the time. It will occur often if events follow normal patterns of process or procedure. The event is repeatable and less sporadic. Auditor/regulator likely to identify issue with minimal audit activity. Process performance failures evident to trained auditors or regulators. | Major and widespread decline in services and quality of life within the community. Infrastructure and communication network severely damaged. Distribution of foods and other basic services paralyzed and can lead to a "a state of primordial chaos and disruption." | Sea-level rise in New York City has averaged 3 centimeters per decade (total of 33.5 centimeters since 1900), nearly twice the observed global rate of 1.3 to 1.8 centimeters per decade over a similar time period. Projections for sea-level rise in New York City are 28 to 53 centimeters by the 2050s, 46 to 100 centimeters by the 2080s, and could reach as high as 1.8 meters by 2100. Health impacts from exposure to extreme weather events include direct loss of life, increases in respiratory and cardiovascular diseases, and compromised mental health. Rising temperatures over the coming century are projected to increase the number of heat-related deaths that occur in Manhattan. The health impacts of Hurricane Sandy varied across the city considerably due to local effects of storm and tidal surges; differing housing types; the degree to which energy, water, and/or transportation infrastructure was disrupted; and the underlying health and resilience factors of the affected population. Vulnerable groups include the old and the very young; women; those with preexisting physical, mental, or substance-abuse disorders; residents of low-income households; members of disadvantaged racial/ethnic groups; workers engaged in recovery efforts; and those with weak social networks (New York City Panel on Climate Change, 2015). |

*(Continued)*

**TABLE A.2 (CONTINUED)**

Researched Megacities Climate Change Vulnerabilities and Health Risks Impacts

| Megacity | Level of Confidence | Level of Likelihood | Level of Consequence | Climate Health-Related Risks Projected Impact |
|---|---|---|---|---|
| Los Angeles | Medium–high confidence (>70%) | Likely to occur; sometimes (occurred infrequently). The activity or event expected to occur 50%–99% of the time. It will occur often if events follow normal patterns of process or procedure. The event is repeatable and less sporadic. Auditor/regulator likely to identify issue with minimal audit activity. Process performance failures evident to trained auditors or regulators. | Major and widespread decline in services and quality of life within the community. Infrastructure and communication network severely damaged. Distribution of foods and other basic services paralyzed and can lead to a chaotic state. | More frequent extreme heat days (defined in Los Angeles as days with daytime maximum temperatures above 92°F) present the greatest potential health threat. High temperatures can lead to dehydration, heat exhaustion, and even deadly heat stroke.<br><br>New research has found Los Angeles County has the largest number of residents in the state that will be exposed to extreme heat and be at greatest risk of heat-related health problems due to factors such as other preexisting health conditions.<br><br>While much improved in recent years, Los Angeles still suffers from some of the worst air pollution in the nation, and our warming climate will make it more challenging to improve or even maintain air quality in the region. Particulate pollution is also a significant problem for Los Angeles; combustion of fuels in cars, trucks, trains, and ships emits tiny particles of hydrocarbons, metals, and other toxics that, when inhaled, can cause serious health problems.<br><br>Particulate pollution is projected to worsen as wildfires, which generate large amounts of smoke and ash, become more frequent and intense (Pacific Institute, 2012). |

*(Continued)*

**TABLE A.2 (CONTINUED)**

Researched Megacities Climate Change Vulnerabilities and Health Risks Impacts

| Megacity | Level of Confidence | Level of Likelihood | Level of Consequence | Climate Health-Related Risks Projected Impact |
|---|---|---|---|---|
| Rio de Janeiro | Very high confidence (>90%) | Likely to occur many times (occurred frequently). It will be continuously experienced unless action is taken to change events. Therefore, it can be expected to occur in most circumstances; more than 75% chance of occurring. | Severe and widespread decline in services and quality of life within the community. Severe, semipermanent, and widespread loss of environmental amenity and likelihood of irrecoverable environmental damage. The region would be seen as very unattractive, moribund, and unable to support its community. | Surface flooding in low-lying areas and landslides on hilly terrain are expected to increase with predictions of more heavy rainfall events. Major flood events in 1967, 1988, and 2010 illustrate the potential for significant impacts, with 200–300 deaths attributed to each event and up to 20,000 people left homeless. The city's topography and climate combine with socioeconomic factors and spatial patterns of urban development to exacerbate vulnerability to floods and landslides. The poorest populations live in informal favelas, often built on steep hillsides. The stripping of hillside forest cover, poor building practices, inadequate drainage infrastructure, and unregulated development contribute to increased vulnerability. Storm events combined with predicted sea-level rise are likely to exacerbate coastal erosion, tidal flooding, and contribute to drainage problems in low-lying areas of the city. |

*(Continued)*

**TABLE A.2 (CONTINUED)**

Researched Megacities Climate Change Vulnerabilities and Health Risks Impacts

| Megacity | Level of Confidence | Level of Likelihood | Level of Consequence | Climate Health-Related Risks Projected Impact |
|---|---|---|---|---|
| | | | | Climate change is expected to exacerbate a number of existing threats to human health and well-being rather than to introduce new health effects (Costello et al., 2009). The health of city dwellers is expected to be affected in the following ways: direct physical injuries and deaths from extreme weather events such as tropical cyclones and other major storms with high winds, storm surges, intense rainfall that leads to flooding, or ice storms that damage trees and overhead structures and produce dangerous transport conditions; illnesses resulting from the aftermath of extreme weather events that destroy housing, disrupt access to clean water and food, and increase exposure to biological and chemical contaminants; waterborne diseases following extended or intense periods of rainfall, ground saturation, and floods and saline intrusion due to sea level rise; foodborne diseases resulting from bacterial growth in foods exposed to higher temperatures; illnesses and deaths from the expanded range of vectorborne infectious diseases; respiratory illnesses due to worsening air quality related to changes in temperature and precipitation resulting in the formation of smog; morbidity and mortality, especially among the elderly, small children, and people whose health is already compromised, as a result of stress from hotter and longer heat waves, which are aggravated by the urban heat island effect; malnutrition and starvation among the urban (and rural) poor who have reduced access to food as a result of drought-induced shortages and price rises; uprooting and migration of populations negatively affected by climate events to areas that are unable to provide the services they need (Barata et al., 2014). |

*(Continued)*

**TABLE A.2 (CONTINUED)**

Researched Megacities Climate Change Vulnerabilities and Health Risks Impacts

| Megacity | Level of Confidence | Level of Likelihood | Level of Consequence | Climate Health-Related Risks Projected Impact |
|---|---|---|---|---|
| Beijing | Very high confidence (>90%) | Likely to occur many times (occurred frequently). It will be continuously experienced unless action is taken to change events. Therefore, it can be expected to occur in most circumstances; more than 75% chance of occurring. | Severe and widespread decline in services and quality of life within the community. Severe, semipermanent, and widespread loss of environmental amenity and likelihood of irrecoverable environmental damage. The negative impacts of climate change on Beijing are increasingly apparent. Consequently, the region would be seen as very unattractive, moribund, and unable to support its community. | Beijing has been struck regularly by urban flooding, and a notable event occurred in July 2012, where 79 people were reported to have been killed, and economic damages were estimated at 10 billion Yuan ($1.6 bn) (Flood Risk Management in China, 2015). Beijing's air pollution levels, as represented by ambient particulate matter concentration, have remained flat for the past 3 years, actually getting slightly worse from 2009 to 2010. This is consistent with results reported in mid-2010. Beijing's air quality still does not meet China's own air quality standard, and PM2.5 measurements are still nearly 6 times worse than WHO-recommended particulate matter targets (Roffo, 2013). Heat-related health effects tend to become more serious at higher temperatures. Multimodel ensembles in probabilistic climate projections indicated cardiovascular mortality could increase by an average of 18.4%, 47.8%, and 69.0% in the 2020s, 2050s, and 2080s under RCP 4.5, respectively, and by 16.6%, 73.8%, and 134% in different decades respectively, under RCP 8.5 compared to the baseline range (Li et al., 2015). |

**TABLE A.3**

Adaptive Capacity and Risk Strategies

| Megacity | Total Population (millions)[a] | Adaptive Capacity and Risk Management Strategy in Place | Health Impacts | Critical Services Interruption | Likelihood Score | Consequence Score |
|---|---|---|---|---|---|---|
| London | 7.61 | Adaptive capacity and risk management strategies in place are enforced uniformly and are not coordinated and tailored with governance policies and initiatives to cope with climate change. Rising population and water demand along with an aging supply system are causing incidents of droughts in the megacity, and the Environment Agency is projecting hotter summers that will put pressure on the infrastructure systems and present a high health risk to their population. | Increased level medical attention (2 weeks' to 3 months' incapacity). | Medium term temporary suspension; backlog cleared by additional resources. | 4 | 3 |

(Continued)

**TABLE A.3 (CONTINUED)**

Adaptive Capacity and Risk Strategies

| Megacity | Total Population (millions)[a] | Adaptive Capacity and Risk Management Strategy in Place | Health Impacts | Critical Services Interruption | Likelihood Score | Consequence Score |
|---|---|---|---|---|---|---|
| Mumbai | 18.84 | Lack access to services that would facilitate adaptation. It needs to assess the megacity's overall risk context, livelihood assets base, and assess the enabling environment. It needs an appropriate assessment of current vulnerability and future climate risks. As Mumbai's population is highly susceptible to climatic impacts, the megacity must conduct an analysis of current vulnerability, which includes the representative vulnerable groups. Narrating potential future vulnerabilities in connection with the present vulnerabilities. Comparing vulnerability under diversified socioeconomic conditions, climate variability, and adaptation responses. | Multiple severe health crises/injury or death. | Indeterminate prolonged suspension of work; nonperformance. | 4 | 5 |

*(Continued)*

**TABLE A.3 (CONTINUED)**

Adaptive Capacity and Risk Strategies

| Megacity | Total Population (millions)[a] | Adaptive Capacity and Risk Management Strategy in Place | Health Impacts | Critical Services Interruption | Likelihood Score | Consequence Score |
|---|---|---|---|---|---|---|
| Lagos | 11.70 | Lacks access to services that would facilitate adaptation. The megacity needs to undertake a detailed and methodology-proven sectoral analysis (e.g., coastal and marine environment, agriculture, water resources, land management, human health) of Lagos' vulnerability to climate change and propose realistic adaptation response strategies and initiatives (programs and projects at all levels, including community-based activities. | Multiple severe health crises/injury or death. | Indeterminate prolonged suspension of work; nonperformance. | 5 | 5 |
| New York City | 18.65 | New York City's ambitious $19.5 billion climate plan seeks to adapt to higher temperatures, sea level rise, and extreme weather impacts. Two-hundred fifty initiatives to reduce the city's vulnerability to coastal flooding and storm surge. New York's climate preparedness strategy includes measures to protect public health from the increasing threat of extreme heat, flood events, and risk of infectious diseases due to extreme weather events caused by climate change. | Increased level medical attention (2 weeks' to 3 months' incapacity). | Medium term temporary suspension; backlog cleared by additional resources. | 4 | 4 |

*(Continued)*

**TABLE A.3 (CONTINUED)**

Adaptive Capacity and Risk Strategies

| Megacity | Total Population (millions)[a] | Adaptive Capacity and Risk Management Strategy in Place | Health Impacts | Critical Services Interruption | Likelihood Score | Consequence Score |
|---|---|---|---|---|---|---|
| Los Angeles | 12.22 | The megacity adaptation component of its plan is a high-level screening analysis, designed to identify some of the most important services and assets that are likely to be affected by climate impacts. The plan outlines options for ensuring that these services and assets continue to function as the climate changes. Nevertheless, heat waves, drought, wildfires, and water supply scarcity will be common factors for the system to collapse. The poorest neighborhoods will be the most affected as the adaptation program is not apt, as of today, to cope with climate change impacts and the security of the energy supply system, health system, and basic services due to its aging infrastructure and governance inefficiencies. | Increased level medical attention (2 weeks' to 3 months' incapacity). | Medium term temporary suspension; backlog cleared by additional resources. | 4 | 3 |

*(Continued)*

**TABLE A.3 (CONTINUED)**

Adaptive Capacity and Risk Strategies

| Megacity | Total Population (millions)[a] | Adaptive Capacity and Risk Management Strategy in Place | Health Impacts | Critical Services Interruption | Likelihood Score | Consequence Score |
|---|---|---|---|---|---|---|
| Beijing | 10.85 | China and in this case Beijing are making big efforts toward mitigation and adaptation strategies with concrete actions. Conversely, actions to offset climate change have initially been slow and uncoordinated. High population density and intensified economic activities are exacerbating environmental issues and climatic impacts on public health. Furthermore, researchers emphasized that the lack of funding for climate adaptation, and the overall absence of a national plan for climate adaptation, has resulted in uncoordinated actions at local government level. | Severe health crisis (incapacity beyond 3 months). | Prolonged suspension of work; additional resources required; performance affected. | 3 | 4 |

*(Continued)*

**TABLE A.3 (CONTINUED)**

Adaptive Capacity and Risk Strategies

| Megacity | Total Population (millions)[a] | Adaptive Capacity and Risk Management Strategy in Place | Health Impacts | Critical Services Interruption | Likelihood Score | Consequence Score |
|---|---|---|---|---|---|---|
| Rio de Janeiro | 11.62 | An adaption strategy is in place and the megacity is actively seeking ways to develop and improve the existing strategies. However, the poor living in informal settlements (favelas) are often driven to reside in dangerous locations due to their inability to pay rent elsewhere, and in order to make a living. Wisner et al. (2003) argue that if residing in a hillside "slum" will lead to economic opportunities; people will choose to live there almost regardless of the disaster risk. Therefore, economic well-being and resilience must be developed conjunctively. Poor urban planning and governance are still a challenge and the urban poor will suffer the most, stressing the habitat and health systems to a collapse. | Severe health crisis (incapacity beyond 3 months). | Prolonged suspension of work; additional resources required; performance affected. | 4 | 4 |

[a] "The 2006 population figures are based on censuses carried out between 2000 and 2005 and adjusted to take account of average annual population changes." Available: www.citymayors.com.

(Adapted from IPCC, 2012: Managing the Risks of Extreme Events and Disasters to Advance Climate Change Adaptation. A Special Report of Working Groups I and II of the Intergovernmental Panel on Climate Change. Field, C.B., V. Barros, T.F. Stocker, D. Qin, D.J. Dokken, K.L. Ebi, M.D. Mastrandrea, K.J. Mach, G.-K. Plattner, S.K. Allen, M. Tignor, and P.M. Midgley (eds.). Cambridge University Press, Cambridge, UK, and New York, NY, USA, 582 pp.)

**TABLE A.4**

Likelihood Score

| Consequence Score | | Population Affected | Reasons for Categorization |
|---|---|---|---|
| 5 | Catastrophic | >1,000,000 | • High death toll<br>• Significant asset destruction or other financial loss<br>• Catastrophic long-term environmental/public health harm<br>• Total cessation of operations |
| 4 | Major | 50,000–1,000,000 | • Multiple serious injuries and death<br>• Acquired illness resulting in hospitalization for more than four weeks<br>• Loss of asset or other financial loss<br>• Significant long-term environmental harm |
| 3 | Moderate | 500–50,000 | • Serious injury or illness resulting in hospitalization for 3 days to 4 weeks<br>• Loss of asset or other financial loss greater than $50,000 up to $1 million<br>• Significant release of pollutants with midterm recovery<br>• Total cessation of operations for less than 2 days |
| 2 | Minor | 50–500 | • Injury or illness resulting in time away from workplace or less than 3 days in hospital<br>• Loss of asset or other financial loss from $5,000 to $50,000<br>• Minor transient environmental harm<br>• Minor disruption to services |
| 1 | Negligible | <50 | • No injuries or minor injuries requiring first aid<br>• Minor loss of asset or other financial loss of less than $5,000<br>• Brief pollution but no environmental harm<br>• No disruption to service |

## Obtaining the Consequence Score

The consequence score qualitatively assesses the risks of each of the identified impacts to determine their relative priority for leading actions to address those risks. This approach is a formalized system that enables the assessment of risks to identify hazards and vulnerabilities of the system in place, populations, health impacts, and so on, when lack of resources and information neglect the implementation of a fully quantitative method (Table A.5).

It is important to consider the future consequences of an impact in order to proactively strategize a plan to combat climate change. It has to be defined as the extent to which megacities consider the potential future outcomes operating under their current condition and the extent to which they are affected by those potential impacts.

**TABLE A.5**

The Consequence Score

| | Likelihood | | | | |
|---|---|---|---|---|---|
| Consequence | Rare (1) | Unlikely (2) | Possible (3) | Likely (4) | Almost Certain (5) |
| Catastrophic (5) | 5 | 10 | 15 | 20 | 25 |
| Major (4) | 4 | 8 | 12 | 16 | 20 |
| Moderate (3) | 3 | 6 | 9 | 12 | 15 |
| Minor (2) | 2 | 4 | 6 | 8 | 10 |
| Negligible (1) | 1 | 2 | 3 | 4 | 5 |

## Obtaining the Risk Score

The risk is the likelihood of harm occurring together with an indication of how serious that harm could be (Table A.6).

Risk rating = Likelihood score × Consequence score

Using Table A.5, find the risk rating (1–25). The risk rating (1–25), see Table A.7, corresponds to a risk score (1–4) as given in Table A.6. Note this in the table.

As described earlier, the potential project health impact can be evaluated through a qualitative and/or quantitative risk rank analysis, and assist in mitigation and adaptation measures. The framework applies to public health events that require an immediate response and are potentially caused by

**TABLE A.6**

The Risk Score

| Risk Score | Risk | Description |
|---|---|---|
| 1 | Very low risk | Local investigation where appropriate. |
| 2 | Low risk | Contributory factor(s) to be identified; discuss with city's management the need for any changes in current and future adaptation strategy, practice, policies, procedures, education, or training. |
| 3 | Moderate risk | Report incident immediately to manager/head of department. Identify contributory factors. Discuss at the city's governance meeting the action plans to be implemented. Assess existing strategies and monitor centrally. |
| 4 | High risk | Report incident immediately to manager/head of department. Inform risk manager. Full investigation to be undertaken including interview with staff and identification of root causes. Action plans to be monitored and reported to central government. Put in place a continuity management system to operate under catastrophic levels of disruption. |

**TABLE A.7**

Megacities' Risk Ratings

| Megacity | Risk Rating |
|---|---|
| London | 15 |
| New York City | 16 |
| Los Angeles | 12 |
| Mumbai | 20 |
| Lagos | 25 |
| Rio de Janeiro | 16 |
| Beijing | 15 |

*Note:* Risk rating = Likelihood score × Consequence score

more than one hazard. It also classifies each potential health effect (very low, low, moderate, and high risk) and they are ranked in order of highest concern by health impact.

The megacity's risk assessment team should decide on a series of initiatives related to their development of a strategy to address climate change impacts on public health. Therefore, leaders must define the scope of the assessment and ensure that all the relevant information is collected. The World Health Organization (WHO) developed a manual to guide rapid risk assessment of acute public health risks from any type of hazard in response to requests from member states. It aims at reducing or preventing disease in affected populations, and minimizes negative social and economic consequences that exacerbate the population vulnerabilities.

Additional benefits include

- Defensible decision making
- Implementation of appropriate and timely control measures
- More effective operational communication
- More effective risk communication
- Improved preparedness (World Health Organization, 2012)

As urban dwellers and local governments will face more intense climate impacts and be forced to cope with rising incidents of disasters, risk assessments become an important part of the strategy to reduce risks, loss, and damage. Using the risk frameworks is an important tool to create a stronger, more resilient community. This methodology to determine the nature and extent of risks by analyzing potential hazards and evaluating existing conditions of vulnerability answers the fundamental question: What would happen if a hazard event occurred in my area (USAID, 2010)? Assessing risk based solely on past events does not provide a comprehensive current state of risks faced by the megacity. Urbanization and rapid population growth are creating high-risk-prone urban areas. Risk management and strategic planning help to improve understanding of climate change patterns as well as their effects on human health. These changes can now be better understood and scenarios for the future developed that allow the policy community to identify adequate strategies for response and adaptation (Kovats et al., 2003).

---

## Rapid Risk Assessment

Risk assessment is a crucial step in managing public health risks. Rapid risk assessment is a systematic process that enables a risk management process during an acute public health event, and assists decision making and operations to improve processes and deliver efficiency while making the best use of the limited time available (Rechel and McKee, 2014). Strategic planning and preparation to identify threats ensure that potential risks are categorized for better understanding in how to address, assess, and manage outcomes.

There are important steps to take to be well prepared and to anticipate any risk that threatens a megacity's functionality of operation and public health:

- Develop evidence-based protocols and guidance as tools to appropriately respond to climatic events and public health risks
- Institute concrete, defined procedures for identifying sources of key information for rapid risk assessment and identify the efficient use of those procedures
- Identify and maintain lists of named individual experts
- Ensure well-trained personnel

The level and severity of climate change impacts on human health depends on many factors that would transpire through pathways of varying complexity, scale, and directness, and with different timing. The actual health impacts, however, are not uniform across countries and regions (Rechel and McKee, 2014). Taking into consideration the changeable nature of the climatic system and impacts, during a catastrophic event, a rapid, concise, and efficient response is crucial to save lives and minimize health risks and damage to the megacity's infrastructure. Responders and well-trained personnel must assess risks in a timely manner and make decisions based on priorities to reduce hazards—all within minutes to hours. It is important to have a prior evaluation of science-based risk assessments for an accurate identification of health impacts to address the potential adverse effects on human health and make projections of the extent and duration of these effects. Megacities' use of the rapid risk assessment tool will be both understandable and applicable to the emergency management community, and would be a valuable asset during any climate disaster (Whelan et al., 2006).

# References

Aagaard-Hansen, J., and Chaignat, C. L. (2010). *Neglected tropical diseases: Equity and social determinants.* World Health Organization.

Abeygunawardena, P. et al. (2003). *Poverty and climate change: Reducing the vulnerability of the poor through adaptation.* World Bank.

Ackerman, F., and Stanton, E. (2011). *The last drop: Climate change and the southwest water crisis.* Stockholm Environment Institute-U.S. Center.

Ackerman, F. et al. (2009). *The economics of 350: The benefits and costs of climate stabilization.* Economics for Equity and the Environment Network.

Adelekan, I. (2010). Vulnerability of poor urban coastal communities to flooding in Lagos, Nigeria. *Environment and Urbanization—International Institute for Environment and Development (IIED)*, 22(2), 2–4.

Adger, W. N. (2006). Vulnerability. *Global Environmental Change*, 16, 268–281.

Aerts, J. C. J. H., Botzen, W. J. W., Emanuel, K., Lin, N., de Moel, H., and Michel-Kerjan, E.O. (2014). Evaluating flood resilience strategies for coastal megacities. *Science*, 344, 473–475.

Ajai, A., Arya, S., Dhinwa, P. S., Pathan, S. K., and Ganesh Raj, K. (2009). Desertification/land degradation status mapping in India. *Current Science*, 97(10): 1478–83.

Al-Moneef, M. et al. (2001). Sector costs and ancillary benefits of mitigation. In *Climate change 2001; Working Group III: Mitigation* (Chapter 9).

Alankar. (2015). *India's megacities and climate change: Explorations from Delhi and Mumbai.* Climate Change and Cities. Brighton, UK: STEPS Centre.

Allwinkle, S., and Cruickshank, P. (2011). Creating smarter cities: An overview. *Journal of Urban Technology*, 18(2), 1–16.

American Lung Association. (2016). *State of the Air.* Retrieved from http://www.lung.org/assets/documents/healthy-air/state-of-the-air/sota-2016-full.pdf.

American Meteorological Society Policy Program. (2014). *A prescription for the 21st century: Improving resilience to high-impact weather for healthcare facilities and services.*

Anderson, G. B., and Bell, M. L. (2012). Lights out: Impact of the August 2003 power outage on mortality in New York, NY. *Epidemiology*, 23(2), 189.

Anderson, T., and Shattuck, J. (2012). Design-based research: A decade of progress in education research? *Educational Researcher*, 41(1), 16–25.

Antonovsky, A. (1996). The salutogenic model as a theory to guide health promotion. *Health Promotion International*, 11(1), 11–18.

Appalachian State University, Department of Physics and Astronomy. (2015). *The inverse square relationship.* Boone, NC: Appalachian State University.

Arkema, K. K., Guannel, G., Verutes, G., Wood, S. A., Guerry, A., Ruckelshaus, M., Kareiva, P., Lacayo, M., and Silver, J. M. (2013). Letter: Coastal habitats shield people and property from sea-level rise and storms. *Nature Climate Change*, 3, 913–918.

Arrhenius, E., and Waltz, T. W. (1990). *The Greenhouse Effect: Implications for economic development.* Washington, DC: World Bank Discussion Papers.

Artuso, F., Chamard, P., Piacentino, S., Sferlazzo, D. M., De Silvestri, L., di Sarra, A., Meloni, D., and Monteleone, F. (2009). Influence of transport and trends in atmospheric $CO_2$ at Lampedusa. *Atmospheric Environment, 43*(9), 3044–3051.

ARUP/C40 Cities Climate Leadership Group. (2011). *Climate action in megacities: C40 cities baseline and opportunities.*

Australian Standard (AS) 8000-2003. (2003). Good governance principles: The responsibilities of shareholders.

Baker, J. (2008). *Urban poverty: A global view.* The International Bank for Reconstruction and Development/The World Bank.

Baker, J. (2011). Climate change, disaster risk, and the urban poor. *The International Bank for Reconstruction and Development/The World Bank,* 7–11.

Baker, J. (Ed.). (2012a). *Climate change, disaster risk, and the urban poor: Cities building resilience for a changing world.* Washington, DC: World Bank.

Baker, J. (2012b). Interview by C. Marolla. Climate change impacts on health: The urban poor in the world's megacities.

Baker, J. (2012c). Opening new finance opportunities for cities to address pro-poor adaptation and risk reduction. In J. Baker (Ed.), *Climate change, disaster risk, and the urban poor: Cities building resilience for a changing world* (pp. 122–123). Washington, DC: The World Bank.

Baliga, S. and Maskin, E. (2003). Mechanism design for the environment. In K.G. Mäler, J. Vincent (Eds.), *Handbook of Environmental Economics* (pp. 306–324). Elsevier Science: North Holland.

Barata, M. (2013, January 8). Interview by C. Marolla. Rio de Janeiro climate change adaptation and mitigation strategies.

Barata, M. et al. (2014). Climate change and human health in cities. *Climate Change and Cities: First assessment report of the Urban Climate Change Research Network,* 179–214. doi:10.1017/cbo9780511783142.013.

Barata, M., Confalonieri, U. et al. (2011). Mapa da vulnerabilidade da população do Estado do Rio de Janeiro aos impactos das mudanças climáticas nas áreas social, saúde e ambiente. Relatório de pesquisa.

Barata, M., Ligeti, E., De Simone, G., Dickinson, T., Jack, D., Penney, J., Rahman, M., and Zimmerman, R. (2011). Climate change and human health in cities. In C. Rosenzweig, W. Solecki, S. A. Hammer, and S. Mehrotra (Eds.), *Climate change and cities: First assessment report of the Urban Climate Change Research Network* (pp. 183–217). Cambridge: Cambridge University Press.

Barnett, J. (2001). *Security and climate change.* Tyndall Centre Working Paper No. 7. University of Canterbury, New Zealand.

Barrett, K. R. (1999). Ecological engineering in water resources: The benefits of collaborating with nature. *Water International, 24*(3), 182–188.

Barrett, P. (1999). *Principles and Better Practices.* Corporate governance in commonwealth authorities and companies: A corporate governance framework.

Barrett, P. (2002). *Achieving Better Practice Corporate Governance in the Public Sector.* International Quality and Productivity Centre Seminar.

Bartlett, S. et al. (2012). *The state of the world's children 2012.* New York: United Nations Children's Fund.

Batagan, L. (2011). Smart cities and sustainability models. *Informatica Economica, 15*(3).

Bates, B. C., Kundzewicz, Z. W., Wu, S., and Palutikof, J. P. (Eds.). (2008). *Climate Change and Water.* Technical Paper of the Intergovernmental Panel on Climate Change, IPCC Secretariat, Geneva. p. 210.

Batty, M. (2008). Cities as complex systems: Scaling, interactions, networks, dynamics and urban morphologies. In R. A. Meyers (Ed.), *Encyclopedia of complexity and systems science* (pp. 1041–1071). Berlin: Springer.

Beatley, T. (2010). *Biophilic cities: Integrating nature into urban design and planning.* Washington, DC: Island Press.

Beatley, T., and Newman, P. (2013). Biophilic cities are sustainable, resilient cities. *Sustainability,* 5(8): 3328–3345.

Beatty, M. E., Phelps, S., Rohner, C., and Weisfuse, I. (2006). Blackout of 2003: Public Health Effects and Emergency Response. *Public Health Reports,* 121(1), 36–44.

Beddington, J. et al. (2012). *Achieving food security in the face of climate change: Final report from the Commission on Sustainable Agriculture and Climate Change.* Copenhagen, Denmark: CGIAR Research Program on Climate Change, Agriculture and Food Security (CCAFS).

Beggs, P., and Bambrick, H. (2006). Is the global rise of asthma an early impact of anthropogenic climate change? *Ciênc. saúde coletiva,* 11(3), 745–752.

Belissent, J. (2011). *The core of a smart city must be smart governance.* Cambridge, MA: Forrester Research, Inc.

Bell, M. et al. (2007). Climate change, ambient ozone, and health in 50 US cities. *Climate Change,* 82, 61–76.

Benson, C., and Clay, E. J. (2004). *Understanding the economic and financial impacts of natural disasters.* Washington, DC: World Bank.

Bergemann, D., and Morris, S. (2012). Robust mechanism design: An introduction. In *Robust Mechanism Design: The role of private information and higher order beliefs* (pp. 1–48). Hackensack, NJ: World Scientific Publishing.

Bernstein, A., and Myers, S. (2011). *Climate change and children's health.* Center for Health and the Global Environment, Harvard Medical School.

Betts, A. (2012). *Extreme weather and climate change.* Alan Betts Atmospheric Research.

Biao, Z. (2011). The climate change, water crisis and forest ecosystem services in Beijing, China. In J. Blanco and H. Kheradmand (Eds.), *Climate change: Socioeconomic effects* (pp. 116–126). InTech.

Biesbroek, G. R. et al. (2009). The mitigation–adaptation dichotomy and the role of spatial planning. *Habitat International,* 33, 230–236.

Bilsborough, D. (2015). *The biophilic city: Can it improve economic prosperity?* Curtin University of Technology. PB-CUSP Alliance-Parson Brinckerhoff.

Blake, R. et al. (2011). Urban climate: Processes, trends, and projections. In C. Rosenzweig et al. (Eds.), *Climate Change and Cities: First Assessment Report of the Urban Climate Change Research Network* (pp. 43–81). Cambridge: Cambridge University Press.

Bleich, A., Gelkopf, M., and Solomon, Z. (2003). Exposure to terrorism, stress-related mental health symptoms, and coping behaviors among a nationally representative sample in Israel. *Journal of the American Medical Association,* 290(5), 612–620. http://doi.org/10.1001/jama.290.5.612.

Boeser, S. M. (2005). Geographic information systems. In *Encyclopedia of Coastal Science,* (pp. 472–474). Dordrecht, Netherlands: Springer.

Börgers, T., Krähmer, D., and Strausz, R. (2015). *An introduction to the theory of mechanism design.*

Bosello, F. et al. (2009a). *An analysis of adaptation as a response to climate change.* Copenhagen: Copenhagen Consensus Center on Climate.

Bosello, F. et al. (2009b). *An analysis of adaptation as a response to climate change.* Ca' Foscari University of Venice, Department of Economics Research Paper Series no. 26.

Braconnot, P. et al. (2007). Understanding and attributing climate change. In *Climate change 2007: The physical science basis. Contribution of Working Group I to the fourth assessment report of the Intergovernmental Panel on Climate Change* (pp. 727–729).

Brauch, H. et al. (2002). *Climate change and conflict prevention.* Federal Ministry for the Environment, Nature Conservation and Nuclear Safety.

Bridges, T. S. et al. (2015). *Use of natural and nature-based features (NNBF) for coastal resilience.* Army Corps ERDC SR-15-1.

Browning, E. K., and Zupan, M. A. (1998). *Microeconomics theory and applications.* Hoboken, NJ: John Wiley & Sons.

Brugmann, J. (2012). Financing the resilient city. *Environment and Urbanization,* 24(1), 215–232.

BS 0:2011. (2011). A standard for standards: Principles of standardization. British Standards Institute.

BS 65000:2014. (2014). Guidance on organizational resilience. British Standards Institute.

Buonaiuto, F. et al. (2011). *Responding to climate change in New York State: The climAID integrated assessment for effective climate change adaptation in New York State.* The New York State Energy Research and Development Authority.

Burdett, R., and Sudjic, D. (2011). *Living in the endless city* (pp. 1–4). London: Phaidon Press.

*Business continuity plan, city of Lincoln, Nebraska.* (2009). City of Lincoln Personnel Policy.

Butler, C. (2012). *Climate change, public health and conflict.* National Centre for Epidemiology and Population Health Research School of Population Health. ANU College of Medicine, Biology and Environment.

Campbell-Lendrum, D., and Corvalán, C. (2007). Climate change and developing-country cities: Implications for environmental health and equity. *Journal of Urban Health: Bulletin of the New York Academy of Medicine,* 84(Suppl. 1), 109–117.

Campbell-Lendrum, D. et al. (2007). Global climate change: Implications for international public health policy. *Bulletin of the World Health Organization,* 85(3), 235–237.

Caragliu, A., Del Bo, C., and Nijkamp, P. (2009). *Smart cities in Europe.* 3rd Central European Conference In Regional Science—CERS, 2009. https://www.inta-aivn.org/images/cc/Urbanism/background%20documents/01_03_Nijkamp.pdf.

Caragliu, A., Del Bo, C., and Nijkamp, P. (2011). Smart cities in Europe. *Journal of Urban Technology,* 18(2), 65–82.

Carbon Disclosure Project. (2011). *Global report on C40 cities.*

Carcavallo, R. U. (1999). Climatic factors related to Chagas disease transmission. *Memorias do Instituto Oswaldo Cruz,* 94(1), 367–369.

Carolan, M. (2004). Ontological politics: Mapping a complex environmental problem. *Environmental Values,* 13(4), 497–522.

Centers for Disease Control and Prevention. (2014). Climate Effects on Health. https://www.cdc.gov/climateandhealth/effects.

Centre for Sustainable Community Development. (2014). *What is sustainable community development?* Simon Frazer University, Burnaby, British Columbia, Canada.

Chorppath, A. K., and Alpcan, T. (2011). *Learning user preferences in mechanism design.* IEEE Conference on Decision and Control and European Control Conference, Orlando, FL, December 12–15.

Chourabi, H. et al. (2012). 45th Hawaii International Conference. HICSS '12 Proceedings of the 2012 45th Hawaii International Conference on System Sciences. IEEE Computer Society, Washington, DC. ISBN: 978-0-7695-4525-7.

Christensen, C. (1997). *The innovator's dilemma: When new technologies cause great firms to fail.* Boston: Harvard Business School Press.

Christensen, C., Horn, M., and Johnson, C.W. (2011). *Disrupting class.* New York: McGraw-Hill.

City of Los Angeles. (2007). *Green LA: An action plan to lead the nation in fighting global warming.*

Clarke, E. H. (1971). Multipart pricing of public goods. *Public Choice,* 11:17–33.

Clemente, J. (2013). *Guide to the study of intelligence: Medical intelligence* (2nd ed., vol. 20). Falls Church, VA: Association of Former Intelligence Officers.

Climate Impacts Group; King County, Washington; and ICLEI (2007). *Preparing for climate change: A guidebook for local, regional, and state governments.*

CNA Corporation. (2007). *National security and the threat of climate change.* Retrieved from http://www.cna.org/sites/default/files/news/FlipBooks/Climate%20 Change%20web/flipviewerxpress.html.

CNA Military Advisory Board. (2014). *National security and the accelerating risks of climate change.* Alexandria, VA: CNA Corporation.

Cohen, B. (2006). Urbanization in developing countries: Current trends, future projections, and key challenges for sustainability. *Technology in Society, 28,* 63–80.

Colle, B. et al. (2008). New York City's vulnerability to coastal flooding. *Bulletin of the American Meteorological Society, 89*(6), 829–841.

Commonwealth Secretariat. (2009). *Commonwealth health ministers' update 2009.*

Confalonieri, U. et al. (2007a). State of health in the world. In M. L. Parry, O. F. Canziani, J. P. Palutikof, P. J. van der Linden, and C. E. Hanson (Eds.), *Climate change 2007: Impacts, adaptation and vulnerability. Contribution of Working Group II to the fourth assessment report of the Intergovernmental Panel on Climate Change* (pp. 393–394). Cambridge: Cambridge University Press.

Confalonieri, U. et al. (2007b). Human health. In M. L. Parry, O. F. Canziani, J. P. Palutikof, P. J. van der Linden, and C. E. Hanson (Eds.), *Climate change 2007: Impacts, adaptation and vulnerability. Contribution of Working Group II to the fourth assessment report of the Intergovernmental Panel on Climate Change* (pp. 391–431). Cambridge: Cambridge University Press.

Confalonieri, U. et al. (2009). Public health vulnerability to climate change in Brazil. *Climate Research, 40*(1), 175–186.

Confalonieri, U., and Margonari de Souza, C. (2015). *Climate Change and Adaptation of the Health Sector: The Case of Infectious Diseases.* Virulence 6(6): 554. http://www .ncbi.nlm.nih.gov/pmc/articles/PMC4720270/.

Conitzer, V., and Sandholm, T. (2007). *Incremental mechanism design.* National Science Foundation and Carnegie Mellon University.

Consensus Energy. (2012). *Goal of energy efficiency audit and upgrade.*

Costa, A., and Kallick, B. (2008). *Learning and leading with habits of mind: 16 essential characteristics for success.* Alexandria, VA: Association for Supervision and Curriculum Development.

Costa Ferreira, L. et al. (2011). Governing climate change in Brazilian coastal cities: Risks and Strategies. *Journal of US-China Public Administration, 8*(1), 51–65.

Costello, A. et al. (2009). Managing the health effects of climate change. *The Lancet, 373*(9676), 1693–1733.

Courtney, H. et al. (2000). *Strategy under uncertainty.* McKinsey & Company.

Cousineau, M. (2009). *Health and health care access in Los Angeles County.* University of Southern California, Keck School of Medicine.

Covenant of Mayors for Climate and Energy. (n.d.). Retrieved from http://www .covenantofmayors.eu/index_en.html.

Cox Downey, D. (2015). *Cities and Disasters.* CRC Press. Taylor & Francis Group. doi: 10.1201/b18964-5.

Cox Jr., L. A. (2002). *Risk analysis foundations, models, and methods.* Springer Science.

Crawford, V. P. (1990). Equilibrium without independence. *Journal of Economic Theory, 50*(1), 127–154.

Creutzig, F., and He, D. (2008). Climate change mitigation and co-benefits of feasible transport demand policies in Beijing. *Transportation Research Part D: Science Direct, 14*, 120–131.

Cui, X. et al. (2009). Threatening of climate change on water resources and supply: Case study of north China. *Science Direct, 248*, 476–478.

Cunningham, J., and Harney, B. (2012). *Strategy and strategists.* Oxford: Oxford University Press.

Dasgupta, P., Hammond, P., and Maskin, E. (1980). On imperfect information and optimal pollution control. *The Review of Economic Studies, 47*(5), 857–860.

De Sherbinin, A., Schiller, A., and Pulsipher, A. (2007). The vulnerability of global cities to climate hazards. *Environment and Urbanization, 19*(1), 39–64.

DeGaetano, A., and Tryhorn, L. (2011). Responding to climate change in New York State: The ClimAID Integrated Assessment for Effective Climate Change Adaptation in New York State. *Annals of the New York Academy of Sciences, 1244*(1). doi:10.1111/j.1749-6632.2011.06331.x.

Delgadillo, R. (2008). Los Angeles Municipal Code. Green Building Program. Los Angeles: Council of the City of Los Angeles.

Denhardt, J., and Denhardt, R. (2009). *Organizational resilience and crisis confronting cities.* Navigating the Fiscal Tested Strategies for Local Leaders—Alliance for Innovation.

Department for International Development (DfID). (2005). *Disaster risk reduction, a development concern.* London: DfID.

Department for International Development (DfID). (2011). *Defining disaster resilience: A DfID approach paper.* London: DfID.

DePaul, M. (2012). Climate change, migration, and megacities: Addressing the dual stresses of mass urbanization and climate vulnerability. *Paterson Review of International Affairs, 12*, 145–162.

Dercon, S. (2003). *Poverty, inequality and growth.* Research Department, Agence Francaise De Developement.

DeVriend, H. J., and Van Koningsveld, M. (2012). *Building with nature.* Dordrecht, Netherlands: EcoShape.

Dewaraja, R., and Kawamura, N. (2006). Trauma intensity and posttraumatic stress: Implications of the tsunami experience in Sri Lanka for the management of future disasters. *International Congress Series, 1287*, 69–73. http://doi.org/10.1016 /j.ics.2005.11.098.

Dhainaut, J. et al. (2004). *Unprecedented heat-related deaths during the 2003 heat wave in Paris: Consequences on emergency departments.* BioMed Central Ltd.

Dickson, E. et al. (2011). *Understanding urban risk: An approach for assessing disaster and climate risk in cities.* World Bank.

Diop, S. (2014). *The land/ocean interactions in the coastal zone of west and central Africa.* Cham, Switzerland: Springer.

Dobbs, R. (2011). *Urban world: Mapping the economic power of cities.* New York: McKinsey Global Institute.

Dollemore, D. (2005). *Obesity threatens to cut U.S. life expectancy, new analysis suggests.* U.S. Department of Health and Human Services.

Draft International Standard ISO/DIS 22313. (2011). Societal security—Business continuity management systems—Guidance (pp. 6–8).

Dukas, R. (Ed.). (1998). Cognitive ecology: The evolutionary ecology of information processing and decision making. Chicago: Chicago University Press.

Ebi, K. L., and Semenza, J. C. (2008). Community-based adaptation to the health impacts of climate change. *American Journal of Preventive Medicine, 35*(5), 501–507.

Ebi, K. L. et al. (2005). *Assessing human health vulnerability and public health adaptation to climate variability and change.* World Health Organization.

Ede, A., and Oshiga, K. (2014). Mitigation strategies for the effects of climate change on road infrastructure in Lagos State. *European Scientific Journal, 10*(11).

Ehrenberg, J., and Ault, S. (2005). Neglected diseases of neglected populations: Thinking to reshape the determinants of health in Latin America and the Caribbean. *BioMed Central Public Health, 5,* 119.

Ellsberg, D. (1961). Risk, ambiguity, and the savage axioms. *Quarterly Journal of Economics, 75,* 643–669.

Emanuel, K. (2005). *Are there trends in hurricane destruction? Nature, 438,* 22–29.

Emanuel, K. et al. (2006). A statistical deterministic approach to hurricane risk assessment. *Bulletin of the American Meteorological Society, 87*(3), 299–314.

Environment Agency. (2009). *Flooding in England: A national assessment of flood risk.* Almondsbury, UK: Environment Agency.

Environment Agency. (2012). *TE 2100 Plan: Managing flood risk through London and the Thames Estuary.* London: Environment Agency, Thames Estuary, 2100.

Envis Centre on Human Settlements. (2006). *Monograph on flood hazard in urban area.* School of Planning and Architecture, New Delhi.

Epstein, P. (2001). West Nile virus and the climate. *Journal of Urban Health: Bulletin of the New York Academy of Medicine, 78*(2), 367–371.

Epstein, P., and Ferber D. (2011). *Changing planet, changing health: How the climate crisis threatens our health and what we can do about it.* Berkeley: University of California Press.

Erhun, F., and Keskinocak, P. (2003). *Game theory in business applications.* Stanford University. Retrieved from http://web.stanford.edu/~ferhun/paper/GT_Overview.pdf.

Eriksen, S. H., and Kelly, P. M. (2007). Developing credible vulnerability indicators for climate adaptation policy assessment. *Mitigation and Adaptation Strategies for Global Change, 12*(4), 495–524. doi:10.1007/s11027-006-3460-6.

European Commission. Mayors Adapt—The Covenant of Mayors Initiative on Climate Change Adaptation. Retrieved from http://mayors-adapt.eu/.

European Commission. (2013). *The Clean Air Policy Package.* Retrieved from http://ec.europa.eu/environment/air/clean_air_policy.htm.

European Commission. (2015a). *Eurosurveillance Monthly Release, 10*(7).

European Commission. (2015b). *European agenda on migration*, Brussels, 13.5.2015—COM (2015).

European Environment Agency. (2005). *Number of reported deaths and minimum and maximum temperature in Paris during the heat-wave in summer 2003.*

European Environment Agency. (2012). *Urban adaptation to climate change in Europe—Challenges and opportunities for cities together with supportive national and European policies.* EEA Report no. 2. Copenhagen, Denmark: European Environment Agency.

European Parliament (2014). *Mapping smart city in the EU.*

European Parliament and the Council. (2010). Directive 2010/75/EU of 24 November 2010 on industrial emissions (integrated pollution prevention and control). The European Parliament and the Council Official Journal of the European Union, 17.12.2010. Retrieved from http://eur-lex.europa.eu/legal-content/EN/TXT/?uri=CELEX:32010L0075.

Eurostat. (2014). *Eurostat regional yearbook.*

Evans, M. (2015). *Megacities: Pros and cons—The case against megacities.* Strategic Studies Institute.

Fann, N. et al. (2011). *Estimating the national public health burden associated with exposure to ambient PM2.5 and ozone.* Society of Risk Analysis.

Feiden, P. (2011). *Adapting to climate change: Cities and the urban poor.* International Housing Coalition.

Fellmann, T. (2012). *The assessment of climate change-related vulnerability in the agricultural sector: Reviewing conceptual framework reviewing conceptual frameworks.* Food and Agriculture Organization of the United Nations. Retrieved from http://www.fao.org/docrep/017/i3084e/i3084e04.pdf.

FEMA: Are You Ready? An In-Depth Guide To Citizen Preparedness. (2004). https://www.fema.gov/pdf/areyouready/areyouready_full.pdf.

Fernando, H. J. S., Klaic, Z. B., and McCulley, J. L. (Eds.). (2012). *National Security and human health implications of climate change.* Dordrecht, Netherlands: Springer.

Fernando, H. J. S., Mammarella, M. C., Grandoni, G., Fedele, P., Di Marco, R., Dimitrova, R., and Hyde, P. (2012). Forecasting PM10 in metropolitan areas: Efficacy of neural networks. *Environmental Pollution, 163*, 62–67.

Fifth Urban Research Symposium. (2009). *Cities and climate change.* The World Bank.

Fiksel, J., Bruins, R., Gatchett, A., Gilliland, A., and ten Brink, M. (2014). The triple value model: Systems approach to sustainable solutions. *Clean Technologies and Environmental Policy, 16*, 691–702.

Flood Risk Management in China. (2015). Collaborative Research on Flood Resilience in Urban Areas (CORFU), Science Policy Brief No. 1. Beijing.

Freitas, C. M., Silva, M. E., and Osorio-de-Castro, C. G. S. (2014). A redução dos riscos de desastres naturais como desafio para a saúde coletiva. *Ciência e Saúde Coletiva* (Impresso), *19*, 3628–3628.

Frenk, J., and Gomez-Dantes, O. (2010). Urban health services and health system reforms. In *Urban health: Global perspectives* (pp. 221–234). San Francisco: Jossey-Bass.

Frenk, J. et al. (2010). Health, nutrition, and public policy. *Food and Nutrition Bulletin, 31*(4).

Friel, S. et al. (2011). Urban health inequities and the added pressure. *Journal of Urban Health: Bulletin of the New York Academy of Medicine, 88*(5), 889–890.

Frost and Sullivan. (2010). *Top 20 global mega trends and their impact on business, cultures and society.*

Frumkin, H. et al. (2008). Climate change: The public health response. *American Journal of Public Health, 98*(3), 435–445.

Frumkin, H., McMichael, A. J., and Hess, J. J. (2008). Climate change and the health of the public. *American Journal of Preventive Medicine, 35*(5), 401–402. Retrieved from http://doi.org/10.1016/j.amepre.2008.08.031.

Gangestad, S. W., and Simpson, J. A. (2007). *The evolution of mind: Fundamental questions and controversies.* New York: Guilford Press.

Gardner, E. et al. (2009). *Study of 16 developing countries shows climate change could deepen poverty.* Purdue University.

Ghosh, A., and Kleinberg, R. (n.d.). *Behavioral mechanism design: Optimal contests for simple agents.* Cornell University. Retrieved from https://www.cs.cornell.edu/~rdk/papers/ec14_GK.pdf.

Gilioli, G., and Mariani, L. (2011). Sensitivity of anopheles gambiae population dynamics to meteo-hydrological variability: A mechanistic approach. *Malaria Journal, 10,* 294.

Gill, S. et al. (2007). Adapting cities for climate change: The role of the green infrastructure. *Built Environment, 33,* 115–133.

Girling, C., and Kellett, R. (2005). *Skinny streets and green neighborhoods: Design for environment and community.* Washington, DC: Island Press.

Glaeser, E., and Kahn, M. (2009). The greenness of cities: Carbon dioxide emissions and urban development. *Journal of Urban Economics, 67,* 404–418.

Glenn, W. (2010). *Climate change mitigation: A strategic approach for cities.* Toronto and Region Conservation.

Global Humanitarian Forum. (2009). *The human impact report: Climate change—The anatomy of a silent crisis.* Retrieved from http://www.ghf-ge.org/human-impact-report.pdf.

Goldsmith, W. (2016). *Integrating natural systems with engineering: Decision support to guide sustainable development and climate resilient design* (PhD thesis). Aarhus University, Herning, Denmark.

Goswami, B., Venugopal, V., Sengupta, D., Madhusoodanan, M., and Xavier, P. (2006). Increasing trend of extreme rain events over India in a warming environment. *Science, 314,* 1442–1445.

Graham, S. (2010). *Disrupted cities: When infrastructure fails.* New York: Routledge.

Grainger, A. et al. (2000). Desertification, and climate change: The case for greater convergence. *Mitigation and Adaptation Strategies for Global Change, 5*(4), 361–377.

Grandoni, G., Mammarella, M. C., and Favaron, M. (2010). Climatology of the BRUNT-VAISALA frequency over Milan, Italy. *Geography Environment Sustainability, 1*(3), 16–24.

Greater London Authority. (2006). *London's urban heat island: A summary for decision makers.* London: Greater London Authority.

Greater London Authority. (2011). *Managing risks and increasing resilience: The mayor's climate change adaptation strategy.* The London Climate Change Adaptation Strategy. London: Greater London Authority.

Greater London Authority. (2014). *London regional flood risk appraisal: First review.*

Green, H. K., Andrews, N. J., Bickler, G., and Pebody, R. G. (2012). Rapid estimation of excess mortality: Now casting during the heat wave alert in England and Wales in June 2011. *Journal of Epidemiology and Community Health, 66,* 866–868.

Grieger, T. a, Fullerton, C. S., and Ursano, R. J. (2003). Posttraumatic stress disorder, alcohol use, and perceived safety after the terrorist attack on the pentagon. *Psychiatric Services*, 54(10), 1380–1382. http://doi.org/10.1176/appi.ps.54.10.1380.

Groves, T. (1973). Incentives in teams. *Econometrica*, 41, 617–663.

Gurney K. et al. (2009). High resolution fossil fuel combustion $CO_2$ emission fluxes for the United States. *Environmental Science and Technology*, 43(14), 5535–5541.

Gurney, K. et al. (2012). Quantification of fossil fuel $CO_2$ emissions on the building/street scale for a large U.S. city. *Environmental Science and Technology*, 46(21), 12194–12202.

Haines, A. (2006). Climate change and human health: Impacts, vulnerability and public health. *Journal of the Royal Institute of Public Health*, 120, 585–596.

Haines, A. et al. (2006). Climate change and human health: Impacts, vulnerability, and mitigation. *London School of Hygiene and Tropical Medicine*, 367, 2101–2109.

Haines, A. (2012). Interview by C. Marolla. Climate change impacts on health: The urban poor in the world's megacities.

Hajat, S., Kovats, R. S., Atkinson, R.W., and Haines, A. (2002). Impact of hot temperatures on death in London: A time series approach. *Journal of Epidemiology and Community Health*, 56, 367–372.

Hall, A. et al. (2012a). *Mid-century warming in the Los Angeles region: Part I of the "climate change in the Los Angeles region" project*. UCLA Department of Atmospheric and Oceanic Sciences.

Hall, A. et al. (2012b). *Los Angeles regional temperature projections*. L.A. Regional Temperature Projections Summary of Findings—June 21, 2012.

Hall, A. et al. (2012c). *Mid-century warming in the Los Angeles region*. Climate Change. L.A.

Hallegatte, S., Bangalore, M., Bonzanigo, L., Fay, M., Kane, T., Narloch, U., Rozenberg, J., Treguer, D., and Vogt-Schilb, A. (2016). *Shock waves: Managing the impacts of climate change on poverty*. Washington, DC: World Bank.

Hallegatte, S., Green, C., Nicholls, R., and Corfee-Morlot, J. (2013). Future flood losses in major coastal cities. *Nature Climate Change*, 3, 802–806.

Hamburg, M. (2008). *Germs go global: Why emerging infectious diseases are a threat to America*. Washington, DC: Trust for America's Health.

Hamer, G. (2003). Solid waste treatment and disposal: Effects on public health and environmental safety. *Biotechnology Advances*, 22, 71–79.

Haque, U. (2012). What is a city that it would be "smart"? *City in a Box*, 34, 140–142.

Harpham, T. (2009). Urban health in developing countries: What do we know and where do we go? *Health and Place*, 15(1), 107–116.

Health Protection Agency. (2006). *Rapid evaluation of 2006 heat wave: Epidemiological aspects*.

Health Protection Agency. (2010). *Chemical hazards and poisons report*. Centre for Radiation, Chemical and Environmental Hazards.

Henn, J., Smorodinsky, S., Attfield, K., and Dobson, C., (2015, February 18). Safe States Alliance Disaster Epidemiology SIG Webinar. Retrieved from http://www.safestates.org/?WebinarCASPER.

Heinzerling, L. (2007). *Climate change, human health, and the post-cautionary principle*. Georgetown University Law Center.

Hickman, R., and Banister, D. (2014). *Transport, climate change and the city*. Oxon: Routledge.

Hobfoll, S. E., Watson, P., Bell, C. C., Bryant, R. A., Brymer, M. J., Friedman, M. J., Friedman, M., Gersons, B. P., de Jong, J. T., Layne, C. M., Maguen, S., Neria, Y., Norwood, A. E., Pynoos, R. S., Reissman, D., Ruzek, J. I., Shalev, A. Y., Solomon, Z., Steinberg, A. M., and Ursano, R. J. (2007). Five essential elements of immediate and mid-term mass trauma intervention: Empirical evidence. Psychiatry, 70(4), 283–315; discussion 316–369. http://doi.org/10.1521/psyc.2007.70.4.283.

Hoffman, A. et al. (2011). *Innovations in higher education: Igniting the spark for success.* Lanham, MD: Rowman & Littlefield.

Hollands, R. (2008). Will the real smart city please stand up? *City,* 12(3), 303–320.

Hollnagel, E. et al. (2006). Resilience: The challenge of the unstable. In *Resilience engineering: Concepts and precepts* (pp. 9–17). Burlington, VT: Ashgate.

Hongyuan, Y. (2008). *Global warming and China's environmental diplomacy* (pp. 52–53). Hauppauge, NY: Nova Publishers.

Hoornweg, D. et al. (2011). Cities and greenhouse gas emissions: Moving forward. *International Institute for Environment and Development,* 23(1), 207–213.

Horton, S. et al. (2009). *Climate risk information.* New York City Panel on Climate Change. Accessed September 12, 2016. http://pubs.giss.nasa.gov/abs/ho00500s.html.

Hotez, P. et al. (2012). Chagas disease: "The new HIV/AIDS of the Americas." *PLOS Neglected Tropical Diseases,* 6(5), 1–4.

Houghton, J. T. et al. (2001). Intergovernmental Panel on Climate Change (IPCC): *Climate Change 2001: The Scientific Basis.* Contribution of Working Group I to the Third Assessment Report of the Intergovernmental Panel on Climate Change. Cambridge University Press, Cambridge, United Kingdom and New York.

HSBC. (2013). *Annual Report And Accounts 2013.* HSBC Holdings plc.

Hunter, P. R. (2003). Climate change and waterborne and vector-borne disease. *Journal of Applied Microbiology,* 94, 37S–46S.

Ibish, H. (2012). Was the Arab Spring Worth It?, *Foreign Policy.*

Ikeda, S. et al. (2006). *A better integrated management of disaster risks: Toward resilient society to emerging disaster risks in mega-cities* (pp. 1–21). Tokyo, Japan: TERRAPUB.

Initiative for a Competitive Inner City. (2011). *Creating shared value: Anchors and the inner city.*

Institute for Global Environmental Strategies. (2009). *Global climate change: Albedo.*

Intergovernmental Panel on Climate Change (IPCC), Working Group II. (2001). Human health. In *Climate change 2001: Impacts, adaptation, and vulnerability* (chap. 9). Cambridge, UK: Cambridge University Press.

Intergovernmental Panel on Climate Change (IPCC). (2007). Stakeholder involvement. In *Climate change 2007: Impacts, adaptation and vulnerability. Working Group II contribution to the Fourth Assessment Report of the Intergovernmental Panel on Climate Change.* Cambridge: Cambridge University Press.

Intergovernmental Panel on Climate Change (IPCC). (2012). *Managing the Risks of Extreme Events and Disasters to Advance Climate Change Adaptation* (582 pp.). A Special Report of Working Groups I and II of the Intergovernmental Panel on Climate Change. C. B. Field et al. (Eds.). Cambridge: Cambridge University Press.

Intergovernmental Panel on Climate Change (IPCC). (2014a). *Climate change 2014: Impacts, adaptation, and vulnerability. Contribution of Working Group II to the Fifth Assessment Report of the Intergovernmental Panel on Climate Change.* Cambridge: Cambridge University Press.

Intergovernmental Panel on Climate Change (IPCC). (2014b). *Fifth assessment report (AR5) of the Intergovernmental Panel on Climate Change.*

Intergovernmental Panel on Climate Change (IPCC). (2014c). Retrieved from ipcc -wg2.gov/AR5/images/uploads/WGIIAR5-Glossary FGD.PDF.

International Organization for Standardization (ISO). (2011). *ISO strategic plan: Solutions to global challenges*. ISO Central Secretariat.

ISO 22301: Societal Security—Business Continuity Management Systems. (2012, May 15). Retrieved January 27, 2016, from http://www.iso.org/iso/catalogue _detail?csnumber=50038.

ISO 31000: 2009, Risk management—principles and guidelines. (2009). New ISO standard for effective management of risk.

ISO 31000: Overview and implications for managers. (2009). InConsult, 5.

Jablonsky, D. (1995). *Time's cycle and national military strategy the case for continuity in a time of change*. Carlisle Barracks, PA: Strategic Studies Institute, U.S. Army War College.

Jackson, M. O. (2003). *Mechanism theory*. Humanities and Social Sciences 228-77. California Institute of Technology.

Jacob, D., and Winner, D. (2009). Effect of climate change on air quality. *Atmospheric Environment, 43*(1), 51–63.

Jedwab, R., and Vollrath, D. (2014). Malthusian dynamics and the rise of the poor megacity. New York University, Marron Institute of Urban Management.

Jensen, M. E. (2007). Evidence and implications of recent climate change. *Climate Change, 72*(3), 251–298.

Jerrett, M. et al. (2005). Spatial analysis of air pollution and mortality in Los Angeles. *Epidemiology, 16*(6), 727–736.

Jha, A. K. et al. (2013). *Building urban resilience principles, tools, and practice*. Directions in Development: Environment and Sustainable Development. International Bank for Reconstruction And Development: The World Bank.

Johnson, H. et al. (2004). The impact of the 2003 heat wave on mortality and hospital admissions in England. *Epidemiology, 15*, S126.

Jones, R. N., and Preston, B. L. (2010). *Adaptation and risk management*. Working Paper No. 15, Centre for Strategic Economic Studies, Victoria University, Melbourne.

Kaczorowska-Ireland, A. (2015). *Competition Law in the CARICOM Single Market Economy*. Routledge.

Kaplan, S., and Garrick, B. J. (1981). On the quantitative definition of risk. *Risk Analysis, 1*, 11–27

Karabag, S. (2011). Climate change management approaches of cities. *European Journal of Economic and Political Studies, 4*(1), 4–7.

Karl, T. et al. (2009). *Global climate change impacts in the United States*. U.S. Global Change Research Program. Cambridge, UK: Cambridge University Press.

Kaye, N. (2011). *BME population in London: Statistical analysis of the latest UK census*. Social Policy Research Centre, Middlesex University.

Kazama, S. (2012). A quantitative risk assessment of waterborne infectious disease in the inundation area of a tropical monsoon region. *Sustainability Sci., 7*, 45–54. doi: 10.1007/s11625-011-0141-5.

Kennedy C. et al. (2009a). *Greenhouse gas emission baselines for global cities and metropolitan regions*. Proceedings of the 5th Urban Research Symposium, Marseille, France, 28–30 June 2009.

Kennedy C. et al. (2009b). Greenhouse gas emissions from global cities. *Environmental Science and Technology, 43*, 7297–7302.

Kersten, E. et al. (2012). *Facing the climate gap*. USC Program for Environmental and Regional Equity and University of California, Berkeley.

Khan, O., and Pappas, G. (2011). *Megacities and public health*. Washington, DC: American Public Health Association.

King, D., and Anderson-Berry, L. (2010). *4.4: Societal impacts of tropical cyclones*. Seventh International Workshop on Tropical Cyclones.

Kinney, P. L., Sheffield, P. E., and Weinberger, K. R. (2014). Climate, air quality, and allergy: Emerging methods for detecting linkages. In K. E. Pinkerton and W. N. Rom (Eds.), *Global climate change and public health* (chap. 7). New York: Springer.

Klein, N. et al. (2012). Globalization of Chagas disease: A growing concern in non-endemic countries. *Epidemiology Research International, 2012*.

Knight, K. (2013, January 18). Interview by C. Marolla. Risk management ISO 31000: Implementing ISO 31000 for major cities to combat climate change impacts on the urban poor's health.

Knowlton, K. et al. (2007). Projecting heat-related mortality impacts under a changing climate. *American Journal of Public Health, 97*(11), 2028–2034.

Komninos, N. (2002). *Intelligent cities: Innovation, knowledge systems and digital spaces*. London: Spon Press.

Komolafe, A. et al. (2014). Air pollution and climate change in Lagos, Nigeria: Needs for proactive approaches to risk management and adaptation. *American Journal of Environmental Sciences*, 10412423(10): 412–423.

Kourtit, K., and Nijkamp, P. (2012). Introduction: Smart cities in the innovation age. *Innovation—The European Journal of Social Science Research, 25*(2), 93–95.

Kourtit, K. et al. (2012). Smart cities in perspective—A comparative European study by means of self-organizing maps. *Innovation—The European Journal of Social Science Research, 25*(2), 229–246.

Kovats, R. et al. (2006). Mortality in southern England during the 2003 heat wave by place of death. *Health Statistics Quarterly, 29*, 6–8.

Kovats, S., and Akhtar, R. (2008). Climate, climate change and human health in Asian cities. *Environment and Urbanization, 20*(1), 165–175.

Kovats, S. et al. (2003). *Methods of assessing human health vulnerability and public health adaptation to climate change*. Health and Global Environmental Change Series no. 1. Copenhagen: World Health Organization.

Krämer, A. et al. (2011). *Health in megacities and urban areas*. Heidelberg: Physica-Verlag.

Kundzewicz, Z. et al. (2008). The implications of projected climate change for fresh-water resources and their management. *Hydrological Sciences–Journal–des Sciences Hydrologiques, 53*(1), 5–7.

Kutasi, G. (2010). *Game theory modelling of climate change aspects*. Corvinus University of Budapest, Department of World Economy.

Ladipo, M. et al. (2011). Recurrent flood: Effects of recent climatic variations on the city of Lagos and its environs. In D. W. Pepper and C. A. Brebbia, *Water and society*. Southampton, UK: WIT Press.

Lafferty, K. (2009). The ecology of climate change and infectious diseases. *Ecological Society of America, 90*(4), 888–900.

Laffiteau, C. (2012). *What types of climate change mitigation and adaptation policies are being considered or implemented in Rio de Janeiro?* The University of Texas at Dallas.

Lake, I. et al. (2012). Climate change and food security: Health impacts in developed countries. *Environmental Health Perspectives, 120*(11), 1520–1521.

Leflar, J., and Siegel, M. (2013). *Organizational resilience managing the risks of disruptive events: A practitioner's guide*. Boca Raton, FL: CRC Press.

Legatum Institute. (2015). Legatum Prosperity Index. Retrieved February 6, 2015, from http://www.li.com/programmes/prosperity-index.

Legros, P., and Cantillon, E. (2007). *The Nobel Prize: What is mechanism design and why does it matter for policy-making?* VoxEU.org CEPR's Policy Portal.

Leiserowitz, A. (2005). American risk perceptions: Is climate change dangerous? *Risk Analysis, 25*(6), 1433–1442.

Levine, D. (2015). *Economic and game theory: What is game theory?* Department of Economics, University of California, Los Angeles. Retrieved May 30, 2015, from http://levine.sscnet.ucla.edu/general/whatis.htm.

Li, T. et al. (2015). Heat-related Mortality Projections for Cardiovascular and Respiratory Disease under the Changing Climate in Beijing, China. *Sci. Rep. Scientific Reports,* 5(2015), 11441. doi:10.1038/srep11441.

Li, T. T., Ban, J., Horton, R. M., Bader, D. A., Huang, G., Sun, Q., and Kinney, P. L. (2015). Heat-related mortality projections for cardiovascular and respiratory disease under the changing climate in Beijing, China. *Scientific Reports, 5,* 11441. doi:10.1038/srep11441.

Li, T. T. et al. (2012). Assessing heat-related mortality risks in Beijing, China. *Biomedical and Environmental Sciences, 25*(4), 458–464.

Lin, N. et al. (2010). Risk assessment of hurricane storm surge for New York City. *Journal of Geophysical Research, 115,* 1–10.

Lin, S., Fletcher, B. A., Luo, M., Chinery, R., and Hwang, S.-A. (2011). Health impact in New York City during the Northeastern blackout of 2003. *Public Health Reports* (Washington, D.C.: 1974), *126*(3), 384–393.

Liotta, P. H. (2007). Environmental change and human security: Recognizing and acting on hazard impacts. Proceedings of the NATO Advanced Research Workshop on Environmental Change and Human Security: Recognizing and Acting on Hazard Impacts Newport, Rhode Island.

Liotta, P., and Miskel, J. (2012). *The real population bomb: megacities, global security and the map of the future.* Washington, DC: Potomac Books.

Liu, W. et al. (2009). Urban-rural humidity and temperature differences in the Beijing area. *Theoretical and Applied Climatology, 96,* 201–207.

Loftis, R. L. (2015). Half of Weather Disasters Linked to Climate Change. Retrieved from http://news.nationalgeographic.com/2015/11/151105-climate-weather-disasters-drought-storms/.

Lombardi, P. et al. (2012). Modelling the smart city performance. *Innovation—The European Journal of Social Science Research, 25*(2), 137–149.

Los Angeles County Metropolitan Transportation Authority. (2010). *Greenhouse gas emissions cost effectiveness study.*

Los Angeles County Metropolitan Transportation Authority. (2012). *Climate action and adaptation plan.*

Lovasi, G. S. et al. (2013). Urban tree canopy and asthma, wheeze, rhinitis, and allergic sensitization to tree pollen in a New York City birth cohort. *Environmental Health Perspectives, 121,* 494–500.

Lowe, A. et al. (2009). Ask the climate question: Adapting to climate change impacts in urban regions. *Center for Clean Air Policy,* 9–10.

Lucier, B., and Syrgkanis, V. (2015). *Greedy algorithms make efficient mechanisms.* Proceedings of the Sixteenth ACM Conference on Economics and Computation.

MacCracken, M. et al. (2008). *Sudden and disruptive climate change: Exploring the real risks and how we can avoid them* (pp. 27–28). London: Earthscan.

MacFarlane, A. (1977). Daily deaths in Greater London. *Population Trends, 5*, 20–25.

Mammarella, M. C., Grandoni, G., and Fedele, P. (2011). *Research on environmental management in a coastal industrial area: New indicators and tools for air quality and river investigations.* Armando Publisher.

Mammarella, M. C., Grandoni, G., Fedele, P., Di Marco, R. A., Fernando, H. J. S., Dimitrova, R., and Hyde, P. (2009). Envinnet: A neural network for hindcasting PM10 in urban Phoenix. AMS, Proceedings J20.5.

Mammarella, M. C., Grandoni, G., Fedele, P., Fernando, H. J. S., Di Sabatino, S., Leo, L. S., Cacciani, M., Casasanta, G., and Dallman, A. (2012). New atmospheric pollution indicators and tools to support policy for environmental sustainable development. In *National security and human health implications of climate change* (pp. 191–197). Dordrecht, Netherlands: Springer.

Mammarella, M. C., Grandoni, G., Fedele, P., Sanarico, M., and Di Marco, R. (2005). *Neural networks for predicting and monitoring urban air pollution: The ATMOSFERA intelligent automatic system.* Bardi editore, Accademia Nazionale dei Lincei, XXIII Giornata dell'Ambiente—Qualità dell'aria nelle città italiane.

Marolla, C. (2016). *Megacity's Mechanism Design with Strategic Adaptive Cognition (SAC) Framework.* Scribd Inc. Print.

Maron, A. (2012). *New York City government: Leading by example.* Department of Citywide Administrative Services, Division of Energy Management, City of New York.

Maskin, E. (2008). *Nash Equilibrium and Mechanism Design.* Institute for Advanced Study and Princeton University. https://www.sss.ias.edu/files/papers/econ paper86.pdf.

Maskin, E. (2008). Nash Equilibrium and Mechanism Design. Institute for Advanced Study and Princeton University. https://www.sss.ias.edu/files/papers/econ paper86.pdf.

Maskin, E. (2013). Mechanism design theory [live interview]. Serious Science.

Math S. B., Nirmala M. C., and Kumar N.C. (2015). *Disaster Management: Mental Health Perspective.* 2015. National Center For Biotechnology Information, U.S. National Library of Medicine. https://www.ncbi.nlm.nih.gov/pubmed/26664073.

Mayor of London. (2007). Action today to protect tomorrow: The Mayor's climate change action plan. Greater London Authority. Available: http://www.lowcvp .org.uk/assets/reports/London%20-%20climate%20change%20action%20 plan.pdf.

McGee, R. (2008). An economic and ethical analysis of the Katrina disaster. *International Journal of Social Economics, 37*(7), 546–557.

McHarg, I. (1969). *Design with nature.* Philadelphia: Natural History Press.

McMichael, A. J. (2004). *Climate change and health: Policy priorities and perspectives.* Briefing Paper. Chatham House.

McMichael, A. et al. (2003a). *Climate change and human health: Risks and responses* (pp. 19–23). Geneva, Switzerland: World Health Organization.

McMichael, A. et al. (2003b). *Climate change and human health: Risks and responses* (pp. 133–147). Geneva, Switzerland: World Health Organization.

Megacity challenges. (2006). *A Stakeholder Perspective.* Gareth Lofthouse, Economist Intelligence Unit.

Mehaffy, M. (2008). Mechanism design theory and sustainable urban form: A proposed priority for collaborative research. Paper presented at Academic Session, Congress for the New Urbanism, 2008. Retrieved from https://www.cnu.org /sites/default/files/mehaffy_cnu16.pdf.

Mehrotra S. C., Rosenzweig, W. D., Solecki, C. E., Natenzon, A., Omojola, R., Folorunsho, J. et al. (2011). Cities, disasters and climate risk. In C. Rosenzweig, W. D. Solecki, S. A. Hammer, and S. Mehrotra (Eds.), *Climate change and cities: First assessment report of the Urban Climate Change Research Network* (pp. 15–42). Cambridge: Cambridge University Press.

Miller, H. E., and Engemann, K. J. (2015). Threats to the electric grid and the impact on organizational resilience. *International Journal of Business Continuity and Risk Management, 6*(1), 1–16.

Mills, B. et al. (2012). *Climate change study on the Los Angeles aqueduct system.* Hydrology Futures, LLC-Los Angeles Department of Water and Power (LADWP).

Mistry, V. et al. (2005). *Financial risks of climate change.* Association of British Insurers.

Moncaster, A. M., Hinds, D., Cruickshank, H., Guthrie, P. M., Crishna, N., Baker, K., Beckmann, K., and Jowitt, P. W. (2010). A key issue: Knowledge exchange between academia and industry. *Proceedings of the Institution of Civil Engineers, Engineering Sustainability, 163*(3), 167–174.

Morello-Frosch, R. et al. (2012). *The climate gap: Inequalities in how climate change hurts Americans and how to close the gap.* The Annenberg Foundation, The Energy Foundation, and the William and Flora Hewlett Foundation, 5–11.

Moser, S. C., and Ekstrom, J. A. (2010). A framework to diagnose barriers to climate change adaptation. *Proceedings of the National Academy of Sciences, 107*(51), 22026–22031.

Muehe, D. (2010). Brazilian coastal vulnerability to climate change. *Pan-American Journal of Aquatic Sciences, 5*(2), 181–182.

Nair, K. (2009). *An assessment of the impact of climate change on the megacities of India and of the current policies and strategies to meet associated challenges.* Nansen Environmental Research Centre (India)—Fifth Urban Research Symposium.

Nam, T., and Pardo, T. (2011). *Conceptualizing smart city with dimensions of technology, people, and institutions.* Proceedings of the 12th Annual International Digital Government Research Conference: Digital Government Innovation in Challenging Times, ACM.

National Aeronautics and Space Administration (NASA). (2014). *2014 climate risk management plan: Managing climate risks and adapting to a changing climate.*

National Bureau of Statistics of China. (2008). *China Statistical Yearbook.* Accessed September 12, 2016. http://www.stats.gov.cn/tjsj/ndsj/2008/indexeh.htm.

National Hurricane Center. (2012). Sandy graphics archive. National Oceanic and Atmospheric Administration.

National Oceanic and Atmospheric Administration. (2012). *Post-Tropical Cyclone Sandy.* National Oceanic and Atmospheric Administration, Office of Response and Restoration.

Neria, Y., Nandi, A., and Galea, S. (2008). Post-traumatic stress disorder following disasters: A systematic review. *Psychological Medicine, 38*(4), 467–480. http://doi.org/10.1017/S0033291707001353.

Neira, M. et al. (2010). *Essential public health package to enhance climate change resilience in developing countries.* World Health Organization.

Neria, Y., and Shultz, J. (2012). *Mental health effects of Hurricane Sandy. JAMA, 308*(24), 2571–2572.

Nerlander, L. (2009). *Climate change and health.* Stockholm: Commission on Climate Change and Development.

New York City Department of Environmental Protection. (2013a). *Climate change integrated modeling project: Phase I assessments of impacts on the New York City water supply.*

New York City Department of Environmental Protection. (2013b). *NYC wastewater resiliency plan: Climate risk assessment and adaptation study.*

New York City Panel on Climate Change 2015 Report Executive Summary. (2015). *Annals of the New York Academy of Sciences, 1336*(1), 9–17. doi:10.1111/nyas.12591.

New York City Panel on Climate Change. (2013). *Climate risk information 2013: Observations, climate change projections, and maps.*

NHS Sustainable Development Unit. (2014). *Adaptation to climate change for health and social care organisations: "Co-ordinated, resilient, prepared."*

Nicholls, R. (2004). *Coastal megacities and climate change* (3rd ed., vol. 37, pp. 369–379). Netherlands: Springer.

Nicholls, R. et al. (2007). Coastal systems and low-lying areas. In *Climate change 2007: Impacts, adaptation and vulnerability. Working Group II contribution to the Fourth Assessment Report of the Intergovernmental Panel on Climate Change* (pp. 315–356). Cambridge: Cambridge University Press.

Noah, D., and Fidas, G. (2000). *The global infectious disease threat and its implications for the United States.* Washington, DC: National Intelligence Council.

Norris, F. H., Friedman, M. J., Watson, P. J., Byrne, C. M., Diaz, E., and Kaniasty, K. (2002). 60,000 disaster victims speak: Part I. An empirical review of the empirical literature, 1981-2001. *Psychiatry, 65*(3), 207–239. http://doi.org/10.1521/psyc .65.3.207.20173.

O'Brien, K. and Leichenko, R. (2000). Double exposure: Assessing the impacts of climate change within the context of economic globalization. *Global Environmental Change, 10*(3), 221–323.

O'Neill, M. S., Zanobetti, A., and J. Schwartz (2005). Disparities by race in heat-related mortality in four US cities: The role of air conditioning prevalence. *Journal of Urban Health: Bulletin of the New York Academy of Medicine, 82*(2), 191–197. Retrieved from http://doi.org/10.1093/jurban/jti043.

Odum, H. T. (1996). Scales of ecological engineering. *Ecological Engineering, 6*(1–3), 7–19.

Omoruyi, E., and Kunle, O. (2011). Effects of climate change on health risks in Nigeria. *Asian Journal of Business and Management Sciences, 1*(1), 204–215.

Oppenheimer, M. et al. (2007). The limits of consensus. *Science, 317*(5844), 1505–1506. Retrieved from http://www.sciencemag.org/content/317/5844/1505 .full.

Organisation for Economic Co-operation and Development (OECD). (2004). *OECD principles of corporate governance.*

Organisation for Economic Co-operation and Development (OECD). (2012). *The OECD environmental outlook to 2050: Key findings on climate change.* Retrieved from http://www.oecd.org/env/cc/Outlook%20to%202050_Climate%20Change %20Chapter_HIGLIGHTS-FINA-8pager-UPDATED%20NOV2012.pdf.

Owen, G. (2008). *Game theory* (3rd. ed.). Bingley, UK: Emerald. (Original work published 1969, Saunders)

Pacific Institute. (2012). Social vulnerability to climate change in California. CEC-500-2012-013. Prepared for the California Energy Commission. Oakland, CA. Online at http://www. energy.ca.gov/2012publications/CEC-500-2012- 013/CEC-500 -2012-013.pdf.

Paes, E., and Rosa, R. (2013, January 31). Interview by C. Marolla. Climate change impacts on health: The urban poor in the world's megacities—Rio de Janeiro.

Park, J., Seager, T. P., Rao, P. S. C., Convertino, M., and Linkov, I. (2013). Integrating risk and resilience approaches to catastrophic management in engineering systems. *Risk Analysis, 33*(3), 356–367.

Parry, M. L. et al. (2007a). Methods used and gaps in knowledge. In *IPCC Fourth Assessment Report: Climate Change 2007* (pp. 394–395). Cambridge: Cambridge University Press.

Parry, M. L. et al. (Eds.). (2007b). *Climate change 2007: Impacts, adaptation and vulnerability. Working Group II contribution to the Fourth Assessment Report of the Intergovernmental Panel on Climate Change*. Cambridge: Cambridge University Press.

Parthasarathy, B., Munot, A., and Kothawale, D. (1994). All-India monthly and seasonal rainfall series: 1871–1993. *Theoretical and Applied Climatology, 49*(4), 217–224.

Pascual, R. (2012, December 12). Interview by C Marolla. Los Angeles climate change adaptation and mitigation strategy. Mayor's Office, City of Los Angeles.

Patankar, A. et al. (2010, August). *Mumbai city report*. Presented at International Workshop on Climate Change Vulnerability Assessment and Urban Development Planning for Asian Coastal Cities, Bangkok, Thailand.

Patil, R., and Deepa, T. (2007). Climate change: The challenges for public health preparedness and response—An Indian case study. *Indian Journal of Occupational and Environmental Medicine, 11*(3), 113–115.

Pearson product-moment correlation. (2012). IBM SPSS Statistics.

Pelling, M. (Ed.). (2011). *Megacities and the coast: Transformation for resilience*. King's College London.

Pennell, N. et al. (2010). Reinventing the city to combat climate change. *Strategy+Business*, Autumn 2010, issue 60.

Piel, G. (1997). The urbanization of poverty worldwide. *Challenge, 40*(1), 58–68.

Pigliucci, M. (2001). Phenotypic plasticity: Beyond nature and nurture. Baltimore: Johns Hopkins Press.

Pojasek, R. B. (2013a). *Moving from environmental management to environmental risk management*. Exponent Engineering and Scientific Consulting.

Pojasek, R. B. (2013b). Organizations and their contexts: Where risk management meets sustainability performance. *Environmental Quality Management, 23*(3), Preprint.

Pontificium Consilium de Justitia et Pace. (2013). *Energy, justice and peace*. Ed. LEV.

Pope Francis. (2015). *Encyclical LAUDATO SI*. Vatican Press.

Porter, M. (2015, February 4). *Aligning strategy and project management*. Speech, Stern Speakers (a Division of Stern Strategy Group.

Portier, C. et al. (2010). *A human health perspective on climate change*. Research Triangle Park, NC: Environmental Health Perspectives and the National Institute of Environmental Health Sciences.

Prasad, N. et al. (2009). *Climate resilient cities: A primer on reducing vulnerabilities to disasters*. Washington, DC: World Bank.

Price-Smith, A. (2002). *The health of nations: Infectious disease, environmental change, and their effects on national security and development*. Cambridge, MA: MIT Press.

Prize Committee of the Royal Swedish Academy of Sciences. (2007). Mechanism design theory. Retrieved from http://www.nobelprize.org/nobel_prizes/economic-sciences/laureates/2007/advanced-economicsciences2007.pdf.

Project RAMSES. (2015). *About the City of Rio de Janeiro*. Reconciling Adaptation, Mitigation and Sustainable Development for Cities.

Projected Climate Change and Its Impacts. (2007). Intergovernmental Panel on Climate Change (IPCC): Climate Change 2007: Synthesis Report. https://www.ipcc.ch/publications_and_data/ar4/syr/en/spms3.html.

Purpura, P. P. (2007). Terrorism and homeland security: An introduction with applications. Amsterdam: Butterworth-Heinemann.

Qi, J. et al. (2007). Environmental degradation and health risks in Beijing, China. *Archives of Environmental and Occupational Health, 62*(1), 33–37.

Raes, F., and Seinfeld, J. (2009). New directions: Climate change and air pollution abatement. *Atmospheric Environment, 43*(32), 5132–5133.

Ramin, B., and Svoboda, T. (2009). Health of the homeless and climate change. *Journal of Urban Health: Bulletin of the New York Academy of Medicine, 86*(4), 1–11.

Rana, A. et al. (2011). Trend analysis for rainfall in Delhi and Mumbai, India. *Climate Dynamics, 38*(1), 45–56.

Rana, A. et al. (2014). Impact of climate change on rainfall over Mumbai using distribution-based scaling of global climate model projections. *Journal of Hydrology: Regional Studies, 1*, 107–128.

Ranger, N. et al. (2010). An assessment of the potential impact of climate change on flood risk in Mumbai. *Climatic Change, 104*(1), 139–167.

Rassi, A. et al. (2007). Predictors of mortality in chronic Chagas disease: A systematic review of observational studies. *Circulation, 115*(9), 1101–1108.

Rassi, A. et al. (2010). Chagas disease. *The Lancet, 375*(9723), 1388–1402.

Razafindrabe, B. et al. (2009). Climate disaster resilience: Focus on coastal urban cities in Asia. *Asian Journal of Environment and Disaster Management (AJEDM), 1*(1), 101–116.

Razmi, A. (2016). The macroeconomics of emission permits: Simple stylized frameworks for short-run policy analysis. *Eastern Economic Journal, 42*(1): 29–45.

Rechel, B., and McKee, M. (2014). *Facets of public health in Europe*. Maidenhead, UK: McGraw-Hill Education.

Reeve, A. et al. (2012). *Informing healthy building design with biophilic urbanism design principles: A review and synthesis of current knowledge and research.* Curtin University of Technology—Healthy Buildings 2012—10th International Conference of the International Society of Indoor Air Quality and Climate (ISIAQ). Brisbane, Australia: Queensland University of Technology.

Revi, A. (2008). Climate change risk: An adaptation and mitigation agenda for Indian cities. *Environment and Urbanization, 20*(1), 207–229.

Rio 2016 Organising Committee for the Olympic and Paralympic Games. (2014). *Carbon Footprint Management Report Rio 2016 Olympic and Paralympic Games.*

Risk Management Fundamentals. (2011). Homeland Security Risk Management Doctrine. https://www.dhs.gov/xlibrary/assets/rma-risk-management-fundamentals.pdf.

Rode, P. et al. (2014). *Accessibility in cities: Transport and urban form*. The New Climate Economy–The Global Commission on the Economy and Climate. London: LSE Cities, London School of Economics and Political Science.

Rodriguez, A. et al. (2009). Impact of climate change on health and disease in Latin America (pp. 12–15). Faculty of Medicine, Universidad Central de Venezuela.

Rodriguez-Oreggia, E. et al. (2010). Natural disasters, human development and poverty at the municipal level in Mexico. *Journal of Development Studies*, *49*(3), 422–455.

Roffo, R. (2013). *A Handbook for Managing Air Pollution Risks in Beijing Metropolitan Area*. Risk Management Policy Paper to the Attention of Beijing Municipality Environmental Protection Bureau.

Rollason, V. et al. (2011). *Applying the ISO 31000 risk assessment framework to coastal zone management*.

Rooney, C. et al. (1998). Excess mortality in England and Wales, and in Greater London, during the 1995 heatwave. *Journal of Epidemiology and Community Health*, *52*, 482–486.

Rosenthal, J. et al. (2007). Links between the built environment, climate and population health: Interdisciplinary environmental change research in New York City. *Urban Climate Impacts and Adaptation*, *36*(10), 834–846.

Rosenzweig C., Solecki, W., Romero-Lankao, P., Mehrotra, S., Dhakal, S., Bowman, T., and Ali Ibrahim, S. (2015). *ARC3.2 Summary for City Leaders*. Urban Climate Change Research Network. Columbia University, New York.

Rosenzweig, C. et al. (2009). *Climate risk information*. New York City Panel on Climate Change.

Rosenzweig, C. et al. (2010). Climate change and global city learning from New York. *Environment: Science and Policy for Sustainable Development*, *43*(3).

Rosenzweig, C. et al. (2011a). Climate change and cities: First assessment report of the urban climate change research network (pp. 17–18). New York: Cambridge University Press.

Rosenzweig, C. et al. (2011b). *Responding to climate change in New York State*. Integrated Assessment for Effective Climate Change Adaptation Strategies in New York State—ClimAID.

Royal College of Midwives. (2002). *What is salutogenesis?* Virtual Institute for Birth: Salutogenesis in Support of Normality.

Ruckelshaus, M., McKenzie, E., Tallis, H., Gerry, A., Daily, G., Kareiva, P., Polasky, S., Taylor, R., Bhagabati, N., Wood, S. A., and Bernhardt, J. (2013). Notes from the field: Lessons learned from using ecosystem service approaches to inform real-world decisions. *Ecological Economics*, *115*, 11–21.

Rumbaitis del Rio, C., and Sjögren, R. (2014). *Scaling up urban climate change resilience: Challenges and opportunities for catalytic donor collaboration*. Meeting document. Bellagio: Rockefeller Foundation.

Rydin, Y. et al. (2012). Shaping cities for health: Complexity and the planning of urban environments in the 21st century. *The Lancet*, *379*, 2079–2108.

Sachs, J. D., and McArthur, J. W. (2005). The millennium project: A plan for meeting the millennium. *The Lancet*, *365*, 347–353.

Satterthwaite, D., Huq, S., Reid, H., Pelling, M., and Romero-Lankao, P. (2008). *Adapting to climate change in urban areas: The possibilities and constraints in low- and middle-income nations*. London: International Institute for Environment and Development (IIED).

Schellnhuber, H. J. et al. (2012). *Turn down the heat: Why a 4°C warmer world must be avoided*. Washington, DC: The Potsdam Institute for Climate Impact Research and the World Bank.

Schroeder, H., and Bulkeley, H. (2009). Global cities and the governance of climate change: What is the role of law in cities? *Fordham Urban Law Journal*, *36*(2), 313–359.

Shankar, P. R., and Rao, G. R. (2002). *Impact of air quality on human health: A case of Mumbai City, India*. Retrieved from http://archive.iussp.org/Bangkok2002/S09Shankar.pdf.

Shea, M. K. (2007). Global climate change and children's health. *Pediatrics, 120*(5), 1149–1152.

Shelley, F. (2015). *The world's population: An encyclopedia of critical issues, crisis and ever-growing countries*. Santa Barbara, CA: ABC-CLIO.

Shenzhen Government Online. (2014). Shenzhen, China: Economic power. Retrieved from http://english.sz.gov.cn/economy/.

Shimizu, K., and Hitt, M. (2004). Strategic flexibility: Organizational preparedness to review ineffective decisions. *Academy of Management Executive, 18*(4), 44–59.

Shultz, J. M., and Neria, Y. (2013). Trauma Signature Analysis State of the Art and Evolving Future Directions. *Disaster Health 1*(1): 4–8. doi:10.4161/dish.24011.

Shultz, J. M., Walsh, L., Garfin, D. R., Wilson, F. E., and Neria, Y. (2014). The 2010 Deepwater Horizon Oil Spill: The Trauma Signature of an Ecological Disaster. *The Journal of Behavioral Health Services & Research.* http://doi.org/10.1007/s11414-014-9398-7.

Shultz, J. M., Marcelin, L. H., Madanes, S. B., Espinel, Z., and Neria, Y. (2011). The "Trauma Signature:" Understanding the Psychological Consequences of the 2010 Haiti Earthquake. *Prehospital and Disaster Medicine, 26*(05), 353–366. http://doi.org/10.1017/S1049023X11006716.

Shultz, J. M., Forbes, D., Wald, D., Kelly, F., Solo-Gabriele, H. M., Rosen, A., Espinel, Z., Mclean, A., Bernal, O., and Neria, Y. (2013). Trauma signature analysis of the great East Japan disaster: Guidance for psychological consequences. *Disaster Medicine and Public Health Preparedness, 7*(July), 201–14. http://doi.org/10.1017/dmp.2013.21.

Shultz, J. M., McLean, A., Herberman Mash, H. B., Rosen, A., Kelly, F., Solo-Gabriele, H. M., Kelly, F., Solo-Gabriele, H. M., Youngs Jr., G. A., Jensen, J., Bernal, O., and Neria, Y. (2013). Mitigating flood exposure. *Disaster Health, 1*(1), 30–44. http://doi.org/10.4161/dish.23076.

Shultz, J. M., Garfin, D. R., Espinel, Z., Araya, R., Oquendo, M. A., Wainberg, M. L., Chaskel, R., Gaviria, S. L., Ordoñez, A. E., Espinola, M., Wilson, F. E., Muñoz Garcia, N., Gomez Ceballos, A. M., Garcia-Barcena, Y., Verdeli, H., and Neria, Y. (2014). Internally displaced "victims of armed conflict" in Colombia: The trajectory and trauma signature of forced migration. *Current Psychiatry Reports, 16*(10), 475. http://doi.org/10.1007/s11920-014-0475-7.

Sitathan, T. (2003, May 23). Singapore's economy: SARS gloom and doom. *Asia Times Online.*

Smith, K. R., Woodward, A., Campbell-Lendrum, D., Chadee, D. D., Honda, Y., Liu, Q., Olwoch, J. M., Revich, B., and Sauerborn, R. (2014). Human health: Impacts, adaptation, and co-benefits. In C. B. Field et al. (Eds.), *Climate Change 2014: Impacts, Adaptation, and Vulnerability. Part A: Global and Sectoral Aspects. Contribution of Working Group II to the Fifth Assessment Report of the Intergovernmental Panel on Climate Change* (pp. 709–754). Cambridge: Cambridge University Press.

State of NSW and Office of Environment and Heritage. (2011). *Guide to climate change risk assessment for NSW local government*. Sydney: Office of Environment and Heritage.

State of Queensland, Department of Employment, Economic Development and Innovation. (2009). *Building business resilience: Business continuity planning guide.*

Stephens, C., Satterthwaite, D., and Kris, H. (2008). Urban health in developing countries. In *International encyclopedia of public health* (pp. 452–463). Oxford: Academic Press.

*Storm water drainage*. (2013). Retrieved from http://www.mcgm.gov.in/irj/go/km/docs/documents/MCGM%20Department%20List/City%20Engineer/Deputy%20City%20Engineer%20(Planning%20and%20Design)/City%20Development%20Plan/Storm%20Water%20Drainage.pdf.

Strategy Under Uncertainty. (2000). McKinsey & Company. http://www.mckinsey.com/business-functions/strategy-and-corporate-finance/our-insights/strategy-under-uncertainty.

Suharno, S. (2010). *Cognitivism and its implications on the second language learning*. Fakultas Ilmu Budaya, Universitas Diponegoro, Indonesia.

Sujaul, I. M. et al. (2013). Effect of industrial pollution on the spatial variation of surface water quality. *Am. J. Environ. Sci., 9*, 120–129. doi: 10.3844/ajessp.2013.120.129.

Sussman, F., and Freed, J. (2008). *Adapting to climate change: A business approach*. Arlington: VA: Pew Center on Global Climate Change.

Sverdlik, A. (2011). Ill-health and poverty: A literature review on health in informal settlements. *Environment and Urbanization, 23*(1), 123–155.

Szwarcwald, C. L. et al. (1999). Income inequality and homicide rates in Rio de Janeiro, Brazil. *American Journal of Public Health, 89*(6), 845–850.

Tarnawski, M. (2013). *Regional Implications of the Arab Spring*. Jagiellonian University, Kraków. http://pressto.amu.edu.pl/index.php/ps/article/viewFile/5948/6013.

Temmerman, R. (2000). *Towards new ways of terminology description: The sociocognitive-approach*. Amsterdam: John Benjamins Publishing Company.

Turnbull, M. et al. (2013). Understanding disaster risk reduction and climate change adaptation. In *Toward resilience: A guide to disaster risk reduction and climate change adaptation* (pp. 1–16). United Kingdom: Practical Action Publishing.

Udomratn, P. (2008). Mental health and the psychosocial consequences of natural disasters in Asia. *International Review of Psychiatry* (Abingdon, England), *20*(5), 441–444. http://doi.org/10.1080/09540260802397487.

Ulanowicz, R. E. (2009). *A Third Window: Natural Life beyond Newton*. Templeton Foundation Press, West Conshohocken, Pennsylvania.

*The UN Global Compact-Accenture CEO Study on Sustainability 2013*. (2013). United Nations Global Compact. Accenture. https://www.accenture.com/_acnmedia/Accenture/Conversion-Assets/DotCom/Documents/About-Accenture/PDF/3/Accenture-13-1739-UNGC-Report-Final-FSC3.pdf.

U.S. Agency for International Development (USAID). (2010). *Risk assessment in cities*. Urban Governance and Community Resilience Guides. Retrieved from http://www.preventionweb.net/files/15047_guidebook02lowres1.pdf.

U.S. Agency for International Development (USAID). (2014). Setting a shared strategic direction for health systems strengthening.

U.S. Department of Defense. (2010). *Quadrennial Defense Review Report 2010*. Retrieved from http://www.defense.gov/qdr/images/QDR_as_of_12Feb10_1000.pdf.

U.S. Department of Defense. (2014). *Quadrennial Defense Review 2014*. Retrieved from http://www.defense.gov/pubs/2014_Quadrennial_Defense_Review.pdf.

U.S. Department of Health and Human Services. (2014). *Primary protection: Enhancing health care resilience for a changing climate*.

U.S. Environmental Protection Agency (EPA). (2010). U.S. Senate Committee on Homeland Security and Governmental Affairs. (2006). *Hurricane Katrina: A nation still unprepared*. Washington, DC: US Government Printing Office.

U.S. Environmental Protection Agency (EPA). (2012). *National ambient air quality standards for particle pollution.*

United Nations Department of Economic and Social Affairs/Population Division. (2012). *World Urbanization Prospects: The 2011 Revision.*

United Nations General Assembly Open Working Group on Sustainable Development Goals. (2014). *Health and sustainable development.*

United Nations High Commissioner for Refugees. (2015). *Environment and climate change.* Retrieved from http://www.unhcr.org/pages/49c3646c10a.html.

United Nations Human Settlements Programme (UN-Habitat). (2003). *The challenge of slums: Global report on human settlements.* London: Earthscan.

United Nations Human Settlements Programme (UN-Habitat). (2011a). *Cities and climate change: The impacts of climate change upon urban areas* (p. 67).

United Nations Human Settlements Programme (UN-Habitat). (2011b). *The impacts of climate change* (pp. 65–68).

United Nations Human Settlements Programme (UN-Habitat). (2013). *State of the world's cities 2012/2013: Prosperity of cities.* Nairobi: UN-Habitat.

United Nations Human Settlements Programme (UN-Habitat). (2015). *Urban resilience.* Habitat III Issues Paper No. 15. New York: United Nations.

United Nations Office for Disaster Risk Reduction (UNISDR). (2008). *Linking disaster risk reduction and poverty reduction.* Geneva: UNISDR.

United Nations Office for Disaster Risk Reduction (UNISDR). (2015). *Global assessment report on disaster risk reduction 2015.* Geneva: UNISDR.

United Nations Programme on HIV/AIDS. (2011). United Nations, (1), 6–7.

United Nations, Department of Economic and Social Affairs. (2014). Population facts: A world of cities. No. 2014/2.

URBACT. (2010). *Cities and the economic crisis: A survey on the impact of the economic crisis and the responses of URBACT II cities.* European Programme for Sustainable Urban Development.

Valent, F., Little, D., Bertollini, R., Nemer, L. E., Barbone, F., and Tamburlini, G. (2004). Burden of disease attributable to selected environmental factors and injury among children and adolescents in Europe. *Lancet, 363*(9426), 2032–2039.

Van Ypersele, J. P. (2007). *Climate change and cities: The IPCC case for action.* C40 Large Cities Climate Summit.

Van Ypersele, J. P. (2013, January 16). Interview by C. Marolla. Climate change impacts on health: The urban poor in the world megacities.

Vandermeer, J. H. (2003). *Tropical agroecosystems.* Boca Raton, FL: CRC Press.

Venton, P., and La Trobe, S. (2008). *Linking climate change adaptation and disaster risk reduction.* London: Tearfund and Institute of Development Studies (IDS).

Villaraigosa, A. (2007). *An action plan to lead the nation.* City of Los Angeles, 2–4.

Visser, W., and Brundtland, G. (2009). *Our common future.* The Brundtland Report. World Commission on Environment and Development.

Vlahov, D. et al. (2011). *Urban health: Global perspectives.* San Francisco: John Wiley & Sons.

Vohra, R. V. (2012). Optimization and mechanism design. *Mathematical Programming, 134*(1), 283–303.

Voiland, A. (2012). *Comparing the winds of Sandy and Katrina.* National Aeronautics and Space Administration (NASA).

Walker, D. (2014). *Under the Weather Improving Health, Wellbeing and Resilience in a Changing Climate.* https://www.gov.uk/government/organisations/environment-agency.

Walters, V., and Gaillard, J. C. (2014). Disaster risk at the margins: Homelessness, vulnerability and hazards. *Habitat International*, 44, 211–219.

Wang, J., and Charneides, B. (2005). *Global warming's increasingly visible impacts*. New York: Environmental Defense.

Washburn, D., and Sindhu, U. (2009). Helping CIOs understand "smart city" initiatives. Forrester Research.

Watson, J. et al. (2007). Epidemics after natural disasters. *Emerging Infectious Diseases*, 13(1).

Wells, J. R. (2012). Strategic IQ. San Francisco: Jossey-Bass.

Werrell, C. E., and Femia, F. (2015). *Climate change as threat multiplier: Understanding the broader nature of the risk*. The Center for Climate and Security.

West-Eberhard, M.J. (2003). Developmental plasticity and evolution. New York: Oxford University Press.

Wheeler, K., Lane, K., Walters, S., and Matte, T. (2013). Heat illness and deaths—New York City, 2000–2011 (Cover story). *MMWR: Morbidity and Mortality Weekly Report*, 62(31), 617–621.

Whelan G. et al. (2006). *Rapid Risk Assessment: FY05 Annual Summary Report*. Pacific Northwest National Laboratory. Retrieved January 7, 2016, from https://www.pnl.gov/main/publications/external/technical_reports/PNNL-15697.pdf.

WHO. (2012). Global Report for Research on Infectious Diseases of Poverty. World Health Organization on Behalf of the Special Programme for Research and Training in Tropical Diseases 2012. http://apps.who.int/iris/bitstream/10665/44850/1/9789241564489_eng.pdf.

Wilkinson, P., Smith, K., Davies, M., Adair, H., Armstrong, B., Barrett, M., Bruce, N., Haines, A., Hamilton, I., Oreszczyn, T., Ridley, I., Tonne, C., and Chalabi, Z. (2009). Public Health Benefits of Strategies to Reduce Greenhouse-gas Emissions: Household Energy. *The Lancet*, 374(9705), 1917–929. doi:10.1016/s0140-6736(09)61713-x.

Wilson, E. (1984). *Biophilia*. Cambridge, MA: Harvard University Press.

Wisner, B. (2003). At Risk: Natural Hazards, People's Vulnerability and Disasters. Routledge.

Wolf, T., and McGregor, G. (2013). The development of a heat wave vulnerability index for London, United Kingdom. *Weather and Climate Extremes*, 1, 59–68.

World Bank. (2008). *World Development Report 2009: Reshaping economic geography*. Washington, DC: World Bank.

World Bank. (2010a). *Cities and climate change: An urgent agenda*. Washington, DC: World Bank.

World Bank. (2010b). *Natural hazards, unnatural disasters: The economics of effective prevention*. Washington, DC: World Bank.

World Bank. (2011). *World Development Report 2011: Conflict, security and development*. Washington, DC: World Bank.

World Bank. (2012a). *Climate change adaptation: Strategy*.

World Bank. (2012b). *Improving the assessment of disaster risks to strengthen financial resilience*. Washington, DC: World Bank.

World Bank. (2012c). *Life expectancy at birth*.

World Bank. (2014). *An expanded approach to urban resilience: Making cities stronger*. Washington, DC: World Bank.

World Council on City Data. (2014). *ISO 37120*.

World Economic Forum. (2013). *Global risks 2013: An initiative of the risk response network* (8th ed.). Geneva, Switzerland: World Economic Forum.

World Health Organization (WHO). (2003). *Climate change and human health—Risks and responses.* Geneva, Switzerland: WHO.

World Health Organization (WHO). (2011). *Disaster Risk Management for Health Fact Sheets: Disaster risk management for health.*

World Health Organization (WHO). (2012). *Rapid risk assessment of acute public health events.*

World Health Organization (WHO). (2013a). *Health and Environment Linkages Initiative: Deaths from climate change.*

World Health Organization (WHO). (2013b). *Public health risk assessment and interventions: Typhoon Haiyan Philippines.*

World Health Organization, Department for International Development. (2009). *Summary and policy implications; Vision 2030: The resilience of water supply and sanitation in the face of climate change.*

World Resources Institute (WRI). (2012). Climate Analysis Indicators Tool (CAIT), version 9.0. Washington, DC: World Resources Institute.

World Water Assessment Programme (WWAP). (2012). *The United Nations world water development report: Managing water under uncertainty and risk.* Paris: UNESCO.

Xu, D. et al. (2011). Assessment of the relative role of climate change and human activities in desertification: A review. *Journal of Geographical Sciences*, 21(5), 926–936.

Yarger, H. R., and Barber, G. F. (1997). *The U.S. Army War College methodology for determining interests and levels of intensity.* Adapted from Department of National Security and Strategy, Directive Course 2: War, National Policy and Strategy. Carlisle, PA: U.S. Army War College, 1997.

Yedla, S. (2003). *Urban environmental evolution: The case of Mumbai.* Institute for Global Environmental Strategy (IGES), Japan.

Yi, C., and Cai, J. (2016). Fundamentals of mechanism design. In *Market-driven spectrum sharing in cognitive radio* (pp. 17–34). Cham, Switzerland: Springer International Publishing.

Young, B. H., Ford, J. D., Friedman, M. J., and Gusman, F. D. (2008). Disaster Mental Health Services A Guidebook for Clinicians and Administrators, 1–2.

Zhao, J. (2011). *Climate change mitigation in Beijing, China.* Case study prepared for Cities and Climate Change: Global Report on Human Settlements. United Nations-Habitat.

Zhao, Z. et al. (2011). The impact of climate change on air quality related meteorological conditions in California—Part II: Present versus future time simulation analysis. *Journal of Climate*, 24.

Zhong, W., Xu, Y., Wang, J., Li, D., and Tianfield, H. (2014). Adaptive mechanism design and game theoretic analysis of auction-driven dynamic spectrum access in cognitive radio networks. *EURASIP Journal on Wireless Communications and Networking*, 2014(1), 44.

Zinni, A. (2010). *Climate change finance: Providing assistance for vulnerable countries. Hearing before the Subcommittee on Asia, the Pacific, and the Global Environment of the Committee on Foreign Affairs, House of Representatives, One Hundred Eleventh Congress, second session, July 27, 2010.* Washington, DC: Government Printing Office.

# *Index*

Page numbers followed by f and t indicate figures and tables, respectively.

National Ambient Air Quality
    Standard, 125
National Environmental Policy Act
    (NEPA), 96
National Guard, 69–71
National Health Service, 88
National Infrastructure Protection Plan,
    166
National Oceanic and Atmospheric
    Administration, 188
National security, climate change and, 160
    American security, 165
    conflict, potential for, 160
    Net Zero approach, 167–177
        army's adaptation strategy,
            168–177, 170f
        Net Zero energy, 169, 170f, 171–172
        Net Zero waste, 169, 170f, 175–177,
            177f
        Net Zero water, 169, 170f, 172–175,
            173f
        resource scarcity, 167–168
    redefining, 163–164
    relationship, 161–163, 162t
    repercussions on, megacities'
        challenges, 167
    ultimate security risk, 165–166
    vulnerability and risks, 164
National Security Act, 165
Natural disasters, 160; *see also* Disasters
    epidemics after, 198
    mental health effects of, 143–147
    and poverty, 19–20
Natural hazard, 41
Natural resource
    water, in Los Angeles, 217–218
Natural systems approach, 108
Neglected diseases of neglected
    populations, 131
Net Zero approach, 167–177
    army's adaptation strategy, 168–177
        Net Zero energy, 169, 170f, 171–172
        Net Zero waste, 169, 170f, 175–177,
            177f
        Net Zero water, 169, 170f, 172–175,
            173f
        overview, 168–171, 170f
    climate change and resource scarcity,
        167–168

Net Zero energy, 169, 170f, 171–172
Net Zero waste, 169, 170f, 175–177,
    177f
Net Zero water, 169, 170f, 172–175, 173f
*New England Journal of Medicine*, 131
New Jersey, Hurricane Sandy, 188–189,
    189f
New York City, 183–195
    adaptation
        programs, 190
        strategy, implementation, 190
    climate-resilient water and
        wastewater infrastructure in,
        192–195
    GHG emissions, mitigation
        strategies, 190–192
        building efficiency, 190–191
        clean distributed generation, 191
        cogeneration/CHP, 191
        $CO_2$ reduction, 192
        energy audits and retrofits, 191
        operations and maintenance, 191
        street lighting, 192
        vehicle fleet, 192
        wastewater treatment plants,
            191–192
    hurricanes; *see also* Hurricanes
        destructive path, 186, 187f
    megacities and megacatastrophes,
        187–188
    mitigation programs, 190
    overview, 181–182, 183
    ozone-related health impacts, 184
    rising sea level, 183, 184f
    sudden and disruptive climate
        change, 185
    UHI effect, 184–185
New York City Climate Change
    Adaptation Task Force, 190
New York City Panel on Climate
    Change (NPCC), 190
*New York Times*, 71
New York University Medical Center,
    65
Nitrogen dioxide, in air, 154
Nitrogen oxides, 236
Nongovernmental organizations
    (NGOs), 68, 258
*NYC Wastewater Resiliency Plan*, 194